Statistical Methods for Building Price Data

Derek T. Beeston

LONDON   NEW YORK
E. & F. N. SPON

First published 1983 by
E. & F. N. Spon Ltd
11 New Fetter Lane, London EC4P 4EE
Published in the USA by
E. & F. N. Spon
733 Third Avenue, New York NY10017

© 1983 Derek T. Beeston

Printed in Great Britain by
J. W. Arrowsmith Ltd, Bristol

ISBN 0 419 12270 2 (hardback)
ISBN 0 419 12280 X (paperback)

This title is available in both hardbound and paperback editions. The paperback edition is sold subject to the condition that it shall not, by way of trade or otherwise, be lent, re-sold, hired out, or otherwise circulated without the publisher's prior consent in any form of binding or cover other than that in which it is published and without a similar condition including this condition being imposed on the subsequent purchaser.
All rights reserved. No part of this book may be reprinted, or reproduced or utilized in any form or by any electronic, mechanical or other means, now known or hereafter invented, including photocopying and recording, or in any information storage and retrieval system, without permission in writing from the Publisher.

British Library Cataloguing in Publication Data

Beeston, Derek T.
Statistical methods for building price data.
1. Building—Estimates—Statistical methods
I. Title
692´.5´072   TH437

ISBN 0-419-12270-2
ISBN 0-419-12280-X Pbk

Library of Congress Cataloging in Publication Data

Beeston, Derek T.
Statistical methods for building price data.
Bibliography: p.
Includes index.
1. Building—Estimates—Statistical methods.
I. Title
TH437.B36   1983   692'.5'072   82-19523
ISBN 0-419-12270-2
ISBN 0-419-12280-X (pbk.)

Statistical Methods for Building Price Data

Contents

1	**Introduction**	**1**
1.1	Who needs statistics?	1
1.2	Judgement	1
1.3	Techniques	2
1.4	Calculators	2
1.5	Computers	3
2	**Variability**	**5**
2.1	The concept of variability	5
2.2	Handling variability	6
2.3	Variability and price books	6
2.4	Random variability	6
2.5	Samples and populations	8
2.6	Measuring variability	11
	2.6.1 Range	11
	2.6.2 Mean deviation	12
	2.6.3 Standard deviation	13
	2.6.4 Coefficient of variation	15
2.7	Wild data and variability	16
2.8	Quantiles	17
2.9	Comparison of ways of measuring variability	18
2.10	Separating causes of variability	20
3	**Selection and adjustment**	**24**
3.1	Selection of data	24
3.2	Descriptions	25
3.3	Random sampling	26
3.4	Indices and their use	28
	3.4.1 Price indices	29
	3.4.2 Price factors	30
	3.4.3 Changes in the mix	31
	3.4.4 Logarithmic plotting	34
	3.4.5 Cost indices	35

	3.4.6	National cost indices	36
	3.4.7	Formula methods	37
	3.4.8	Ratio of price to cost	37
3.5	Adjustment of data		37
	3.5.1	Adjustment factors	37
	3.5.2	Subjective adjustment	38
	3.5.3	The importance of adjustment	39
3.6	Blending data		39
	3.6.1	Weighting data	40
	3.6.2	Measuring the variability of weighted data	41
3.7	Data banks		41
	3.7.1	Edge-punched cards	42
	3.7.2	Captive punched cards	43
	3.7.3	Optical coincidence cards	44
	3.7.4	Machine-operated punched cards	45
	3.7.5	Computer methods	45
	3.7.6	Data structure	46

4	**Patterns in prices**		**49**
4.1	Distributions		49
4.2	Skewness		49
4.3	Transformations		52
4.4	Logarithmic transformation		52
4.5	Square root transformation		54
4.6	The normal distribution		54
4.7	Tables of the normal distribution		55
4.8	Using the normal curve		57
4.9	Differences in patterns (the χ^2 test)		58

5	**Confidence in estimates**		**64**
5.1	Confidence limits		64
	5.1.1	Confidence limits for the arithmetic mean	64
	5.1.2	Blending estimates of the population standard deviation	68
	5.1.3	Confidence limits for the median	69
	5.1.4	Confidence limits for other parameters	70
	5.1.5	Quoting confidence limits	71
	5.1.6	Asymmetrical confidence limits	71
	5.1.7	Single-sided confidence limits	72
5.2	Deciding sample sizes		72
5.3	Testing significance		74
	5.3.1	Student's t test	75
	5.3.2	Single- or double-tailed probabilities	77
	5.3.3	The null hypothesis and limits of difference	78
	5.3.4	Non-parametric tests	80
	5.3.5	The Mann–Whitney test	80

6	**Data appreciation**	**83**
6.1	Appraising data	83
6.2	Arrangement of raw data	83
	6.2.1 Frequency tables	85
	6.2.2 Histograms	86
	6.2.3 Calculations from frequency tables	88
6.3	Measures of location	89
	6.3.1 Arithmetic mean	89
	6.3.2 Trimmed mean	91
	6.3.3 Calculating the arithmetic mean	91
	6.3.4 Geometric mean	92
	6.3.5 Median	92
	6.3.6 Mode	93
	6.3.7 Weighted mean	95
	6.3.8 Weighted median	96
	6.3.9 Harmonic mean	97
6.4	Rapid appraisal	97
	6.4.1 Examples of rapid appraisal	100
6.5	Watching for changes	105
	6.5.1 Control charts	105
	6.5.2 Cusum charts	108
7	**Tender patterns and bidding strategy**	**110**
7.1	A model for tender patterns	110
7.2	Non-serious tenders and cover prices	112
7.3	Dispersion of tenders	112
7.4	Bidding strategies	114
7.5	The D curve method	115
	7.5.1 The D curve	115
	7.5.2 Micro-climate	117
	7.5.3 Macro-climate	117
	7.5.4 Number of tenders	118
	7.5.5 Updating the population coefficient of variation	120
7.6	Summary of the D curve method	120
7.7	Using the strategy	121
7.8	Example	122
	7.8.1 Preparation	122
	7.8.2 Use	124
7.9	Further development	125
7.10	Testing and tuning	125
8	**Cost models**	**127**
8.1	Types of cost model	127
8.2	Realistic methods	127

	8.2.1	Data banks for realistic methods	129
	8.2.2	Variability in realistic methods	130
8.3	In-place materials methods		130
	8.3.1	Variability in in-place materials methods	132
	8.3.2	Data banks for in-place materials methods	132
8.4	Area-related methods and cost planning		134
8.5	Area-related methods and estimating		135
8.6	Regression methods		139
	8.6.1	Outlying data	139
	8.6.2	Linearity	140
	8.6.3	Correlations	141
	8.6.4	Residuals	141
	8.6.5	Significance of the coefficients	143
	8.6.6	Dummy variables	144
	8.6.7	Stepwise regression	144
	8.6.8	Population drift	144
8.7	Design assumptions		145

9 The accuracy of estimating — 147
9.1 Measuring performance — 147
9.2 Present achievement — 149
9.3 Improving estimating performance — 149
9.4 Helping the contractor — 152
9.5 Computer methods — 153

10 Forecasting — 154
10.1 The need for forecasting — 154
10.2 Principles of forecasting — 154
10.3 Project expenditure forecasting — 154
   10.3.1 Construction programmes — 154
   10.3.2 S-curves — 155
   10.3.3 Planned and actual duration — 158
10.4 Forecasting price movements — 158
   10.4.1 Forecasting by analogy — 159
   10.4.2 Detecting a seasonal pattern — 160
   10.4.3 Seasonal adjustment — 163
   10.4.4 Smoothing — 164
   10.4.5 Adjustment for inflation — 167
   10.4.6 Other adjustments — 167
   10.4.7 Forecasting by projection and adjustment — 167

Tables — 169

Further reading — 172

Index — 173

1 Introduction

1.1 Who needs statistics?

There are many statistics books written for particular groups of readers but, until now, none for those who deal with building price data. These include students, research workers, builders' estimators, estate surveyors, architects and, of course, in many countries, quantity surveyors. Their needs are very different from other students of statistical techniques because of the peculiar nature of building price data and their special position in relation to their clients. They are in the position of being expected to give accurate advice without having the quality of data capable of providing it.

They exercise judgement and privately worry that their results are not really good enough, while having to display confidence to their clients or employer. They sometimes turn to research, hoping for a magic formula, but are always disappointed because they expect too much. This does not mean that there is no hope of improvement and this book shows how small but solid gains can be made.

Many research workers in the field of building prices already use statistical methods, and some practitioners have felt the need to study the subject just so that they can decide whether and how to use the fruits of research. Those whose livelihood depends on their handling of price data are justifiably cautious in their approach to new methods. They will accept only what they can understand and prefer to do their own research, using the research of others to confirm and broaden their own findings.

Those who have learned something of the statistical approach to their work have found that they can improve their cost advice to clients by studying the variability of prices using statistical tools. They have also found they have been able to make selective use of statistically based research work and have quickly realized that there need be no mystery in research methods.

1.2 Judgement

For most practitioners, allowing for variability is a matter of judgement. The development of judgement is the result of experience, and those who have worked for a long time in the field are rightly proud of the ability they have developed. For instance, they have a feeling for how far data can be relied upon and how much to allow for possible error.

This judgement is slowly and painfully built up as a result of making many mistakes and learning from them. The ability is almost impossible to transfer to others new to the work so all have to go through the same long process. Having done so, it is difficult for the experienced practitioner to believe that there is a quicker way which reduces his judgement to numbers and which provides a reliable medium for the rapid transfer of experience. Statistical methods provide this medium and have been developed for the purpose of handling variability and evaluating uncertainty.

1.3 Techniques

It is sometimes said that advanced analytical techniques are not appropriate to data as poor as those found in the world of building prices. There is some reason in this point of view in that precision is impossible so there is no point in working to many places of decimals; but it ignores the fact that statistical techniques are designed to make best use of poor data. In fact, the worse the data the more need there is for a statistical approach.

Data vary in quality but some use can be made of nearly all. With so little reliable data available, not even the least reliable should be completely ignored. Limits can be placed on uncertainty and, in this book, special methods of blending data of different quality have been devised.

Too much must not be expected but statistical methods, appropriately used, quantify uncertainty and reduce it. This can be of great importance and, taken along with the improvement to powers of judgement already described, more than justify the effort expended in studying the methods selected and adapted for inclusion in this book.

Only techniques which have been found useful for building price data have been included and some have been adapted by the author to provide new techniques using the principles of existing methods.

1.4 Calculators

This book is intended for a wide variety of readers. No assumptions have been made about mathematical ability, access to computing aids or about the type or size of problems to be dealt with. Those who have the use of an advanced calculator or a computer will find the methods described enable them to make more powerful use of them, but they are far from necessary.

The most useful type of calculator is one with a log button and one for standard deviation. The latter can be regarded as essential. It is usually labelled with the sign σ. Also useful, but very much less important, is the ability to perform regressions.

When data are being entered for the calculation of standard deviation the machine should respond quickly after each entry of a number so that there will be little danger of entering the next number before the machine is ready to receive it.

An audible or click response to key depression is helpful but some find the heavy pressure associated with the latter to be tiring.

A programming facility is useful, but it must be remembered that the main disadvantage of a calculator compared to a computer is the need to re-enter data for each analysis. If adequate storage is provided a programmable calculator becomes a computer.

A minor point, but one which sometimes arises when checking someone else's calculations, concerns rounding off. Some calculators have an automatic rounding off facility. This is useful but it should be remembered that it always rounds 5 upwards. Thus $2.5/2 = 1.25$ exactly becomes 1.3 when rounded to one place of decimal. Such rounded numbers may be recorded and re-entered later for multiplication and the products totalled. If there are many such numbers there will be an upward bias in the total. It is good practice to round such exact calculations ending in 5 always to the nearer even number. This produces about the same number of downward rounding errors as upward.

1.5 Computers

As well as speeding up the absorption of experience by providing a language for the description of numerical data, the use of well-chosen statistical methods allows large quantities of data to be summarized with little more difficulty than small quantities. This is especially important now that more data are becoming available for analysis by computer. Anyone faced with a large computer data bank would have to use statistical summarizing methods to deal with large blocks of data taken out of the bank. Statistical methods described in this book have been found to be the most appropriate.

Apart from selecting data from a large bank, for the actual analysis the only technique included in this book for which a computer is of great assistance is multiple regression analysis, so no one need be deterred by not having access to advanced aids to calculation. A major aim of this book is to improve the rapid appreciation of data, and for this purpose complex analysis is unnecessary. The development of statistically based judgement arises from the study of how data behave using simple statistical tools applied to a wide variety of data. When conclusions have to be drawn from data, the possible loss from too elementary an analysis is less important than the danger of being so intent on the application of an advanced method that sight is lost of the meaning and applicability of the results.

For the selection, handling and simple analysis of fairly large amounts of data there is little good statistical software mounted on any but the largest computers. Readers with access to these will need to consult their software libraries to see what is available. A few owners of mini-computers, especially in education, have mounted some statistical software but there is at present little available for micro-computers. Users of these will probably write their own selection and analysis programs as they find the need to supplement the inadequate software provided.

A practical way of obtaining access to the statistical libraries of large computers is to open an account with a time-sharing computer bureau. Time-sharing systems are normally accessed from teletype terminals which simply send signals along the telephone line, though they may have some buffer storage capability. A recent development by some time-sharing bureaux is to offer connections to users' micro-computers. If the system's program library includes good data handling and analysis programs, this should make a powerful combination. It should provide economy for the initial stages of data handling and analysis combined with the availability of advanced techniques when required, without the need to re-enter data. When choosing a micro-computer it would be wise to consider the possibility of connection to a time-sharing network. If contact is made with a few time-sharing computer bureaux the choice of micro-computer can be made so as to keep such an option open.

Costs of using time-sharing systems can be surprisingly low but can rise rapidly if great care is not taken, especially in file handling and storage. For the small user, storage of data on micro-computer discs would be cheaper than the cheapest form of storage on a time-sharing system. For the very large user the time-sharing computer bureaux offer bargains in storage costs but even he should consider local storage for at least some of his data files.

Points to look for when choosing a bureau or combination of micro-computer and bureau are the ease and economy with which data can be amended, updated, selected, rearranged and combined. This is not easy to judge from manuals and price books, so costed tests on full-sized files are required. Suitable files can usually be called up by the bureau without the customer having to input large amounts of data simply for test purposes.

Some statistical library programs produce tables and graphs of adequate quality for use without transcription. Elimination of this source of error, coupled with the saving of time and typing effort, is so important that it should be a major criterion when choosing a system.

The methods used by statistical library programs are sometimes criticized. For instance, the multiple regression program sometimes makes it difficult for the user to keep as much control of variable selection as he would wish. The random number generator may not pass the most stringent tests of randomness. However, conclusions drawn from data by users of the programs are extremely unlikely to be affected by such shortcomings in them.

2 Variability

2.1 The concept of variability

The statistical concept of variability is subtle but it is essential to an understanding of the application of statistical methods to building price data.

One starting point is the variability between prices for buildings fitting the same description. An example might be the price per unit floor area of similarly sized office blocks built in the same city in the same year of about the same standard of quality.

Someone who knew nothing of building prices would be surprised at how much they varied and might conclude that, in the light of the circumstances of its contract and construction, the description of the building was too broad and that narrowing it would remove most of the variability.

Anyone who analyses building price data will discover that, no matter to what extent descriptions are narrowed, a disappointing amount of unexplained variability remains. It is this which has held back progress in estimating and set limits to the accuracy with which it is possible to forecast the price at which a contract will be let.

Contractors' estimators working under identical conditions produce surprisingly different estimates from each other. Naturally, estimators working for clients will have less knowledge of builders' costs and must expect that their estimates will vary even more in relation to the tenders which they are trying to forecast. Whether clients' estimators differ much from each other is unimportant except in as much as it increases their variability in relation to tenders.

Variability of prices in relation to estimates could only be eliminated by an agreed method of pricing, by not going ahead unless the price can be made to equal the estimate by negotiation or by only accepting tenders which are close to the estimate. Such procedures would require complete confidence in estimating data and methods while at the same time undermining estimating accuracy by complicating the gathering of the price data required to feed the estimating method. However, it may be that such ways of setting a contract price will one day become normal and the natural variability of prices thereby concealed. Meanwhile, it is important to find ways of measuring the variability while it is in its 'natural state' and coping with it. Statistical methods provide the best approach.

2.2 Handling variability

The usual reaction when faced with variability is first to try to reduce it in some way. When no more reduction is possible what remains is usually ignored. However, sometimes variability makes itself obvious and produces a feeling of unease.

As a result, the usual request to a source of cost analyses is for just one which fits the desired description. The provision of more than one is considered counterproductive and professional judgement is used to select one of them. The rest are discarded, even when their building descriptions are equally appropriate.

The author concludes from discussion with practitioners that the main cause of reluctance to use more than one price analogue is the desire to ignore at the earliest stage the existence of worrying unexplained variability, so that it can be forgotten.

It is a major objective of statistical methods to provide ways of handling variability and using it constructively to measure and improve the reliability of calculations based on data. Far from trying to ignore variability, the statistical method seeks to obtain sufficient data to measure it.

2.3 Variability and price books

Where price books can be used to provide unit prices for measured work, they appear to side-step the problem of variability of price data. In fact they only conceal it.

Standard unit rates are built up from prices for materials, labour and plant. These are variable but their variability is more difficult to measure than data gathered directly from records for a number of buildings. The fact that the variability is difficult to measure does not mean that its effect is less. It does mean that, when attempting to set limits to the possible error in an estimate, the variability of some of the rates may have to be assessed subjectively.

2.4 Random variability

Using 'description' in its widest sense to include the circumstances of construction such as labour costs and market factors, as a description is narrowed so the variability of price data from buildings conforming to the description is reduced. Unfortunately the number of buildings conforming to the description is also rapidly reduced, and is soon reduced to one.

That is what happens in practice but, to take a highly theoretical point of view, consider that all known ways of narrowing the description had been exhausted but that we were still left with a number of apparently identical buildings constructed in apparently identical circumstances. The prices of these apparently identical buildings would not be the same. We might conclude that the variability was caused by some factors and circumstances which we do not know about; perhaps they are known only to the contractors. Whatever the reasons for the variability, if we cannot know them we can do nothing about them. This 'hard-core' variability can be called 'random' as far as we are concerned. A less commonly used but better name for it is 'unassigned' variability.

Somebody else going through the same theoretical exercise using the same data would take into account different factors and circumstances and would arrive at a different amount of unassigned variability.

It may be that this 'hard-core' or 'random' variability is due not only to factors which we know nothing about but also to some which we have chosen to disregard. This may be because the work entailed in studying them is more than seems justified. Somebody else going through the same exercise may be ignorant of, or choose to disregard, a different set of factors from the ones which we omitted. In that case the hard-core variability measured by them will be different.

Thus there is no definite amount of unassigned, or random, variability inherent in any set of price data. It differs according to how it is being measured or defined. Even the same person may go different distances along the path of narrowing descriptions in different exercises and thus arrive at different amounts of unassigned (i.e. random) variability. In fact the term 'random variability' is often applied even when the description is deliberately not narrowed very much so as to avoid eliminating too many buildings from those which conform to the description. The penalty of a greater amount of random variability may have to be accepted.

This concept of random variability is essential to an understanding of even the most elementary application of statistical methods. It is so important that it is worth taking a little more time to approach it from another point of view.

Instead of narrowing a description, imagine that we can accurately adjust the price for each building for each descriptive factor in turn. In that way we should avoid eliminating any buildings. If we could adjust perfectly for all possible factors we should end with all prices equal.

If that were possible we should have assigned all the variability to specific factors. Of course it is impossible to go as far as this. The variability which cannot be assigned or, from choice or practical necessity, is not assigned is called random variability, unassigned variability or, in some statistical analyses, residual variability.

Unless it is borne in mind that random variability is affected by the amount of variability which has been assigned to factors or eliminated by the chosen level of building description, it is easy to fall into the trap of assuming that it is solely inherent in the data.

2.5 Samples and populations

The next important concepts which must be understood, if the approaches and techniques described in this book are to be useful, are those of a sample and a population. These can be most simply dealt with by defining a population as all data, real or imaginary, which conform to a given description; and a sample as a group of data selected from this population. The idea of an imaginary population larger than the real one is dealt with later.

Most samples are intended or believed to represent the population from which they were drawn. There is no way of making certain that this is so without examining the whole population, but if the sample is large enough and is selected randomly it will be found to represent the essential features of the population within limits which can be calculated.

Statistical theory deals almost exclusively with random samples because of the mathematical difficulty in dealing with other types of sample. With the growing use of computer methods this may change. Some of these non-random sampling methods are described below and in specialized textbooks.

It is possible to improve the representativeness of the sample by deliberately selecting cases with certain features so that these features are present in the correct proportions. This tends to produce a better representation of other features which are associated with them.

This is done by breaking the population down into groups according to the chosen feature or features and randomly sampling separately within each group. The proportions sampled from the groups can be the same or can be chosen differently to ensure adequate representation of less common or especially interesting groups and the avoidance of waste which might otherwise be caused by unnecessarily large numbers of very common or uninteresting groups.

In Table 2.1, 1- and 2-storey buildings are far more common than taller buildings. If they are no more interesting there is no point in expending more effort on studying them. Using the same sampling proportions throughout would give a sample in which more than 7/8 of the pitched roof buildings studied would be of 1 or 2 storeys. This could be avoided if the sampling proportion for 1- and 2-storey buildings were much smaller than that for taller buildings.

It is convenient to use simple proportions such as 1%, 5%, 10%, 20%, 25% or 50%. In Table 2.1, 10% has been chosen for the largest groups and 50% for those groups of medium size. The smallest groups are included in full – i.e. 100% sampling. This gives a total sample size of 110. If this was thought to be too large for the resources available for the study, the sampling proportions could be reduced.

If the only objective is the best possible estimate of the population average, the sampling proportions should be very roughly proportional to the variability within the groups, but this rule is not often applicable because variability can seldom be measured before the study has been done.

Separation of the population into groups for the purpose of sampling is called *stratification*. Stratified samples with different sampling proportions in the groups, or strata, must have this deliberate imbalance corrected by rescaling before the strata can be combined.

The example in Table 2.1 shows a convenient way of doing this which includes the calculation of group means. These are usually interesting in themselves. The estimated whole population mean is $13\,758.1 \div 487 = 28.3$.

Table 2.1

Number of storeys (a)	Type of roof (b)	Number in population (c)	Random sampling proportion (d)	Number in sample (approx. $c \times d$)*	Mean annual maintenance expenditure per unit area in sample (calculated from sample) (e)	Estimated total for population (mean × number in population ($c \times e$))
1 or 2	Pitched	221	0.1	22	20.4	4508.4
	Flat	136	0.1	11	43.8	5956.8
3, 4 or 5	Pitched	25	1.0	25	15.1	377.5
	Flat	52	0.5	27	25.7	1336.4
6 or more	Pitched	5	1.0	5	24.9	124.6
	Flat	48	0.5	20	30.3	1454.4
		487		110		13 758.1

* In some methods of random sampling the sampling proportion is only approximately achieved. See Section 3.3.

The proportions sampled and the numbers in the samples are not used in the calculation but it is useful to record them for future reference when the exercise might be repeated.

With most modern calculators the last column would not have to be written down.

With some methods of sampling, the numbers in the population may not be known exactly. In such cases they can be estimated as in Table 2.2.

This time the estimated population mean is $12\,395.0 \div 454 = 27.3$. The difference from the previous example reflects the inaccuracy introduced by having to estimate the numbers in the population.

The idea that a sample can be selected randomly or according to a rule designed to give a representation of the population seldom causes difficulty. The only problems arise in connection with the population.

For instance, there are cases where the difference between a sample and its

Table 2.2

Number of storeys (a)	Type of roof (b)	Random sampling proportion (c)	Number in sample (d)	Estimated number in population (d ÷ c)	Sum of annual expenditures per unit area in sample (e)	Estimated sum for population (e ÷ c)	Mean annual maintenance expenditure in sample (also estimate for population) (e ÷ d)
1 or 2	Pitched	0.1	22	220	448.2	4482	20.4
	Flat	0.1	11	110	481.3	4813	43.8
3, 4 or 5	Pitched	1.0	25	25	378.6	378.6	15.1
	Flat	0.5	27	54	693.3	1386.6	25.7
6 or more	Pitched	1.0	5	5	124.6	124.6	24.9
	Flat	0.5	20	40	605.1	1210.2	30.3
			110	454		12 395.0	

population are difficult to see. The population may seem to coincide with the sample in as much as all available cases have been included in the sample.

When considering variability it is often necessary to think in terms of a larger hypothetical population which could exist if the imagination were stretched far enough. This is the larger imaginary population already mentioned. One way of rationalizing this is to think of buildings, conforming to a chosen description, which may be constructed in the future or elsewhere. This hypothetical population is valid whether or not such buildings are in fact constructed.

With familiarity this concept of a hypothetical population becomes a useful way of viewing all sorts of data. The item or sample of items under consideration is then seen against a background of others conforming to the same description. This background is the population and the item or sample might, by chance, have been any of them. This way of looking at data characterizes the statistical approach as distinct from the deterministic view which sees each item as unique and particular.

A further, and especially difficult, example of a hypothetical population is a population of lowest tenders for one particular, unique contract. There can in fact be only one lowest tender but from the variability of the other tenders it is possible, with certain assumptions, to calculate a notional variability of the lowest tender for such competitions.

Such a concept can be rationalized by imagining the competition being repeated many times with the competitors having no memory of the past. The tenders vary from one competition to another, different contractors win, and a population of lowest tenders is produced of which the one in hand would be a member.

2.6 Measuring variability

Before discussing causes of variability it is necessary to describe the ways of measuring it.

2.6.1 Range

The range is the obvious method of measuring variability and is popularly used by even the least numerate. In such uses the highest and lowest figures are quoted, but statistically speaking only the difference between them can be described as the range.

The range might be expected to suffer from the disadvantage of being very sensitive to wild data. Building price data sometimes contain one or two items which are out of scale with the rest; often due to mistakes. In this book these are called 'wild data' and are certain to increase the range, but surprisingly it is hardly more affected than the more highly regarded measures described later. Of course the range cannot distinguish between patterns of data with bunching at the centre, those more evenly spread and those with a 'hole in the middle'. However, such a variety of patterns seldom needs to be distinguished.

The range has the great advantage of easy calculation so it is often used when the greater accuracy of other methods is not worth the extra effort.

For small samples (fewer than about 6 items) the range is very nearly as good a measure of variability as any other. Only when sample sizes exceed about 15 does the range show up as being markedly inferior.

There is, however, a credibility aspect to consider. Non-statisticians find it difficult to accept that two items from a sample can represent not only the variability of the sample but also that of the population from which it was drawn. They forget that the two items are not just any two items but only have a meaning in relation to the rest of the sample. To this extent they contain information not just about themselves but about the other items as well.

2.6.2 Mean deviation

A better measure of variability than range is the mean deviation. This is based on deviations from the arithmetic mean (commonly called the average – but there are other averages).

The mean deviation is the arithmetic mean of the deviations when deviations above and below the mean are all treated as positive. (If they were not, the mean deviation would always be zero.) Table 2.3 shows how it is calculated.

Ignoring the sign makes the mean deviation difficult to use mathematically. It cannot be represented by a readily manipulated algebraic formula. However, it is easy to appreciate and its sensitivity to wild data is not great.

Table 2.3

	Data	Deviations from mean
	1.1	+0.17
	0.7	−0.23
	0.5	−0.43
	1.3	+0.37
	1.0	+0.07
	0.9	−0.03
	0.8	−0.13
	1.0	+0.07
	1.1	+0.17
Total	8.4	Total (check = approx. 0) +0.03
Number of items	9	Total ignoring signs 1.67
Arithmetic mean	0.93	Mean deviation 1.67/9 = 0.19

For the user who needs to measure variability only for purposes of simple comparison, the mean deviation is quite satisfactory and worth using purely because it is easy to visualize and easy to calculate. A simple calculator with a memory or a constant facility is all that is required.

2.6.3 Standard deviation

The mathematical inconvenience produced in the mean deviation by neglecting the sign is not present in the standard deviation. In this measure of variability the deviations are also measured from the arithmetic mean, but before addition they are squared. Because squares of negative numbers are positive, the difficulty of sign is removed.

The arithmetic mean of these squared deviations is calculated and its square root is taken. This gives the standard deviation.

There is a computational formula which makes it unnecessary to calculate the arithmetic mean of the data or, which is the important thing, the deviations from it. The data values themselves are squared and totalled. The sum of the original data values is squared and the following formula applied.

$$\text{Standard deviation} = \sqrt{\left[\frac{\text{sum of squares} - (\text{square of sum/number of items})}{\text{number of items}}\right]}$$

If it eases the calculation to subtract a constant amount from every item of data this can be done with no effect on the standard deviation. However,

Table 2.4

Data	Squares of data
1.1	1.21
0.7	0.49
0.5	0.25
1.3	1.69
1.0	1.00
0.9	0.81
0.8	0.64
1.0	1.00
1.1	1.21

Total 8.4   Sum of squares 8.30
Number of items 9
Square of sum 70.56

$$\text{Standard deviation} = \sqrt{\left[\frac{8.30 - (70.56/9)}{9}\right]}$$

$$= 0.23$$

multiplying by a constant must be compensated for at the end by dividing the standard deviation by the constant.

The standard deviation is slightly more affected by wild data than the mean deviation but less so than is generally believed. The squaring of deviations is usually said to render the standard deviation markedly sensitive to extreme values and common sense indicates that this should be so. However, a few experiments with actual or fictitious, but likely, data soon show that there is little to choose between range, mean deviation and standard deviation in this respect. An example is given in Section 2.9.

The above methods of calculation are correct for measuring the standard deviation of a group of data. This may sometimes be required for its own sake when studying causes of variability and when comparing with other groups of data. More often, though, the standard deviation of a random sample is calculated so as to provide an estimate of the standard deviation of the population.

The sum of the squared deviations is less when measured from the arithmetic mean than when measured from any other value. The following example illustrates this. The arithmetic mean of the data in Table 2.5 is 11 and the sum of the squared deviations from it is 168. The sums of the squared deviations from 10 and 12, or from any other number, are greater than 168.

Table 2.5

Data	Deviations from mean (=11)	Squares	Deviations from 10	Squares	Deviations from 12	Squares
3	−8	64	−7	49	−9	81
8	−3	9	−2	4	−4	16
9	−2	4	−1	1	−3	9
11	0	0	+1	1	−1	1
12	+1	1	+2	4	0	0
14	+3	9	+4	16	+2	4
20	+9	81	+10	100	+8	64
		168		175		175

When used to estimate the population standard deviation, the sample standard deviation has to be adjusted by a factor known as *Bessel's correction*. This is because the deviations from which the standard deviation is calculated are measured from the sample's own arithmetic mean and not that of the population. Although the sample arithmetic mean is a good estimate of the population arithmetic mean it is unlikely to be exactly equal to it. Therefore, without Bessel's correction, the estimate of the population standard deviation is biased. It tends to be too small.

The size of the adjustment depends on the number of items in the sample (n) and it is made by multiplying the sample standard deviation by $\sqrt{[n/(n-1)]}$. This is Bessel's correction.

To express the adjustment as a formula requires the introduction of a symbol for the sample standard deviation (s) and one for the unknown population standard deviation (σ). Then the best estimate of σ is

$$s \sqrt{\left(\frac{n}{n-1}\right)}$$

σ is the universal symbol for the population standard deviation and is the lower case Greek letter 'sigma'.

Bessel's correction is for bias and makes the adjusted sample standard deviation the best estimate of the population standard deviation. It will be right 'on average' and 'in the long run' but not, of course, every time. It can be described as an unbiased estimator of the population standard deviation.

Similarly, because the arithmetic mean of a random sample is the best estimator of the population arithmetic mean it is said to be an unbiased estimator of the population arithmetic mean.

2.6.4 Coefficient of variation

When populations are being compared with respect to, for example, cost per unit area, it is to be expected that one with an arithmetic mean twice as great as the other will also have a standard deviation about twice as great. So that the variability of populations can be compared when the arithmetic means are very different, the standard deviation is often expressed as a percentage of the arithmetic mean. This percentage is called the coefficient of variation.

In the earlier stratified sampling example the coefficients of variation (c.v.) of the strata can be calculated, as in Table 2.6, by expressing the estimated

Table 2.6

Number of storeys	Type of roof	Sample mean (=estimated population mean)	Estimate of population s.d. (σ)	Estimate of population c.v. (%)
1 or 2	Pitched	20.4	4.41	22
	Flat	43.8	19.00	43
3, 4 or 5	Pitched	15.1	2.89	19
	Flat	25.7	12.77	50
6 or more	Pitched	24.9	4.18	17
	Flat	30.3	11.24	37
Whole population		28.3	11.84	42

population standard deviations (s.d.) as percentages of the corresponding arithmetic means from Table 2.1.

2.7 Wild data and variability

Although the standard deviation and the coefficient of variation are by far the commonest ways of measuring variability, there are other ways whose value in building price work lies in their insensitivity to wild data.

It is the author's opinion that users of building price data are too ready to exclude inconvenient prices from their analyses. An item from one of the extremes of a population has only a small chance of turning up in a random sample, but when it does it can look so different from the rest of the sample that it is hard to believe that it could have come from the same population.

It is tempting to exclude such an item in the belief that either it really belongs to a different population or it is the result of an error. It certainly has a strong influence on the arithmetic mean and a case can be made for excluding or adjusting it for that reason. However, in the measurement of variability, there are better ways out of the difficulty. Corresponding ways of preventing wild data affecting the average are given in Sections 6.3.2 and 6.3.5.

If the wild data are excluded or adjusted, either a large element of subjectivity is introduced or rules have to be formulated which assume more knowledge of the population than is usually available. Such a rule is to exclude data which lie more than so many standard deviations from the mean. When both mean and standard deviation are estimated from the sample this is a dubious procedure. The method can be refined but it is better to avoid the difficulty by using other measures of variability.

2.8 Quantiles

The mean deviation and standard deviation are inflated by the inclusion of wild data but there are other measures, based on quantiles, which are much less so if at all.

If data are placed in ascending or descending order of the variable being examined, it is possible to state a value of the variable which will be exceeded by any chosen percentage of the items. This value is called a *quantile* or *percentile*. It is labelled by the proportion or percentage of items which exceed it or which it exceeds.

The most commonly used quantiles are the *quartiles*. These mark the 25% and 75% points so that a quarter of the items are less than the lower quartile and a quarter are greater than the upper quartile.

The quartiles of the following ten numbers (already arranged in ascending order)

14.2, 15.1, 15.2, 15.9, 16.5, 16.6, 16.7, 18.1, 18.9, 19.2

are 15.2 and 18.1 because, as nearly as can be, a quarter of them (actually one-fifth) lie below 15.2 and above 18.1 and a half (actually two-fifths) lie between them. As nearly as possible the same number of items lies between the quartiles as outside them.

The difference between the quartiles is called the *interquartile range* (IQR) and can be used as a measure of variability. Also used is half the IQR, called the semi-interquartile range (SIQR). In the above example the IQR is 2.9 and the SIQR is 1.45.

Where there are enough data to warrant it (more than about 30 items) it is reasonable to use the *deciles* which mark the 10% and 90% points. The interdecile range is seldom used however, perhaps because with such large samples there is less concern for the sensitivity of the standard deviation to wild data.

For some numbers of items the quartiles fall exactly on two of the items. With 6 items the quartiles are the second and fifth. With 10 items they are the third and eighth. With 14 items they are the fourth and eleventh, and so on. For other numbers of items the choice is more difficult and in consequence is sometimes made unnecessarily roughly, and differently by different people. The following simple method will make it exactly and consistently.

Arrange the data in ascending or descending order and use the following formula to find the 'distance' of the quartiles from each end of the list.

If there are n items this 'distance' is

$$\frac{n}{4} + \frac{1}{2}$$

A few examples will show how to use the formula.

First, one of the simple, exact cases. With 6 items of data

$$\frac{n}{4} + \frac{1}{2} = 2$$

so the lower quartile is the second number from the smaller end of the list and the upper quartile is the second number from the larger end.

For cases where the quartile does not fall exactly on an item the formula gives a fractional answer requiring interpolation, as in the following examples.

For the four numbers 8, 6, 5, 2 (i.e. $n = 4$),

$$\frac{n}{4} + \frac{1}{2} = 1\tfrac{1}{2}$$

which means that the quartiles are half-way between the first and second numbers from each end. The upper quartile is obviously 7. The lower quartile is half-way between 2 and 5, that is, $3\tfrac{1}{2}$. This gives the quartiles $3\tfrac{1}{2}$ and 7. The IQR is the difference between them, i.e. $3\tfrac{1}{2}$, and the SIQR is half this, i.e. $1\tfrac{3}{4}$.

Now the most difficult case where the formula gives a less convenient fraction, for example, where $n = 7$. If the numbers are

$$1, 3, 3, 6, 8, 11, 12$$

$$\frac{n}{4} + \frac{1}{2} = 2\tfrac{1}{4}$$

The lower quartile is obviously 3.

The upper quartile is found by taking a quarter of the difference between the second and third items from the higher end, i.e. $(11 - 8)/4 = \tfrac{3}{4}$, and subtracting it from 11. This gives the upper quartile as $10\tfrac{1}{4}$, the IQR as $7\tfrac{1}{4}$ and the SIQR as $3\tfrac{5}{8}$.

Even when the data are not integers the method is exactly the same. As an exercise, confirm that the quartiles of 2.1, 3.4, 4.0, 4.7, 8.2 are 3.1 and 5.6.

For deciles the formula is

$$\frac{n}{10} + \frac{1}{2}$$

and is applied in the same way.

The general formula to give the 'distance inward' for any quantile is

$$\frac{n}{Q} + \frac{1}{2}$$

where Q is the 'order' of the quantile (e.g. 4 for quartiles and 10 for deciles).

Measures of variability based on quantiles have the disadvantage that, in the past, less attention has been paid to them by theoreticians because they are more difficult and less rewarding to handle mathematically than the standard deviation. Recently their value has been recognized more widely so their theory has been developed and methods of application regularized. The subject heading under which this has been done is 'order statistics' and some statistics books, especially these on distribution-free or non-parametric methods, devote a large section to it.

2.9 Comparison of ways of measuring variability

The following example shows how the various measures respond to changes in variability when the data are 'well-behaved'.

The four groups of data in Table 2.7 have the same arithmetic mean but the variability of each is greater than the one to its left. In the boxes the measures for each group are expressed as ratios to those for the first group.

There is little difference in the response of the various measures. Judged by this crude and rather idealized example, all would be equally useful. The reader would gain valuable experience by similarly experimenting to form his or her own opinion of their properties.

Variability 19

Table 2.7

	A		B		C		D
	7		5		3		1
	9		8		7		6
	10		10		10		10
	10		10		10		10
	11		12		13		14
	13		15		17		19
		B/A		C/A		D/A	
Arithmetic mean	10		10		10		10
Mean deviation	1.3	1.8	2.3	2.5	3.3	3.3	4.3
Estimated population standard deviation	2.0	1.7	3.4	2.4	4.8	3.1	6.2
Range	6	1.7	10	2.3	14	3.0	18
Interquartile range	2	2.0	4	3.0	6	4.0	8
Coefficient of variation (of sample)	18%	1.7	31%	2.4	44%	3.1	57%

The example in Table 2.8 compares the way in which the same measures of variability are affected by increasing 'wildness'. They are calculated for each group of data. The four groups differ only in the biggest item. Once again, the figures in the boxes are the ratios of the values of the measures for each group to their values for the first group.

Table 2.8

	A		B		C		D
	1		1		1		1
	2		2		2		2
	3		3		3		3
	4		4		4		4
	5		10		20		90
		B/A		C/A		D/A	
Arithmetic mean	3		4		6		20
Mean deviation	1.2	2.0	2.4	4.7	5.6	23.3	28.0
Estimated population standard deviation	1.6	2.2	3.5	5.0	7.9	24.8	39.1
Range	4	2.3	9	4.8	19	22.3	89
Interquartile range	$2\frac{1}{2}$	1.5	$3\frac{3}{4}$	2.5	$6\frac{1}{4}$	9.5	$23\frac{3}{4}$
Coefficient of variation (of sample)	47%	1.7	79%	2.5	118%	3.7	175%

The range, mean deviation and standard deviation in this example are affected almost equally by the increasing 'wildness' of the largest item. The interquartile range is less affected and the coefficient of variation even less so.

Of course, insensitivity to wild data is not necessarily a desirable feature of a measure of variability. It is in most cases, but in each application its particular effect must be considered. If a measure of variability is being used for descriptive rather than analytical purposes, as all but the standard deviation usually are, it can be misleading if it does not respond to a wildly high or low item.

A good strategy is to note the presence of wild data but having done so to use a measure of variability which is insensitive to them.

2.10 Separating causes of variability

The standard deviation and the coefficient of variation are the most useful measures for the mathematical study of causes of variability. Using them total variability can be broken down into components due to separate causes.

Data on costs per unit area would generally be less variable if they were composed of only one type of building than if they included several types; so, in a mixed set of data, the factor 'building type' is a cause of variability in the variable 'cost per unit area'. Larger buildings of the same type tend to cost less per unit area than smaller buildings, so the more variability there is in building sizes in a set of data, the more the cost per unit area must be expected to vary. Looking at it another way, if the effect of size on cost is known, even roughly, adjusting prices for the effect should tend to reduce variability.

These approaches are combined in statistical terms in the techniques of analysis of variance (often abbreviated to ANOVA), multiple regression analysis and other multivariate analytical methods. Using these methods the effects of different causes of variability can be separately measured.

Analysis of variance is designed to quantify the amount of variability contributed by each of the causes, called 'factors'. Multiple regression analysis is a procedure for calculating for each factor a multiplier representing its contribution in more directly usable terms.

Multiple regression analysis has found more application than analysis of variance in building cost work because the factors and their multiplier can form an estimating formula. Unfortunately, in their enthusiasm for the technique, some researchers have applied it in misleading ways and read too much into the results obtained from analyses based on data from far too few projects.

The application of multiple regression analysis and analysis of variance is described in Section 8.6 but some results of using analysis of variance are given below to help the reader assess the usefulness of evaluating causes of variability.

The analyses were made by the author using indexed Bills of Quantities from the Government's Property Services Agency (PSA). The aim was to quantify the effects of a few factors fairly accurately. Most analyses go no further than examining a large number of factors and showing that the apparent effects of some of them in the sample were larger than could be explained by chance alone. This leaves very wide limits on the evaluation of the effects. About 100 projects would have been enough for this but to achieve a useful degree of quantification needed about 1000 projects.

The significant factors successfully quantified were building function (e.g. telephone exchange, barracks, etc.), building value (or size) and location.

There must, of course, be many other such factors and work is going on to find and quantify them. The effort required is great and expensive unless the search is narrowed to the influence of factors in very particular circumstances. It is more economical to manipulate designs to which a pricing method has been applied, changing one factor at a time and measuring its effect on price, but less confidence can be placed in the results because of the huge assumptions implicit in such methods. Little variation of one factor is possible without affecting others and assumptions about the effect on others have a strong influence on the resultant prices.

The figures given below are from an analysis of variance of the PSA projects. Although it is not exactly how analysis of variance works, the simplest way of viewing the analysis is to consider that the effect of removing a factor is shown by calculating the coefficient of variation of the price data before and after adjusting them for the factor.

The adjustment multipliers were calculated from the same data, because no more effort was available to apply them to a different set, so they were 'ideal' for those data and would have less effect on a different set. Thus their effects are slightly exaggerated by the figures below.

The price data analysed were from more than 1000 buildings. They were total contract sums, each expressed as a percentage of a standard price for the contract. The standard price was calculated by repricing each Bill of Quantities from a schedule of unit rates. When the price data are costs per unit area similar relativities are found but all coefficients of variation are greater.

	Coefficient of variation (%)
Adjusted only for time	10.9
Also adjusted for location	9.6
Also adjusted for function and value	8.6

The analysis was made using standard deviations but coefficients of variation are used for presentation because of their generality.

After the data had been adjusted for time by dividing them by the appropriate tender price index, the gain from further adjustment was small even though the factors for which the data were adjusted are important. This is disappointing, especially when the slight exaggeration explained above is taken into account.

The reason is that when the factors are operating independently from each other – that is when a change in one does not tend to be associated with a change in another – their effects on total variability are additive by squares.

Table 2.9 shows that, although adjusting data for location produces only a small reduction in the coefficient of variation, it is a very important cause of variability. If the only difference between buildings was that they were in different locations, the coefficient of variation of their price levels would be about 5%. It could possibly be slightly less than indicated by this analysis because some unidentified effects could have been operating which were associated with location and which came out in the analysis as due to location.

The consideration of function and value together was to avoid the complication of their association with each other. This association came out strongly in the analysis but in any case was obvious. Some building functions

Table 2.9

Factor	Coefficient of variation (%) (CV)	$(CV)^2$
Total	10.9	119
− Location	−5.2	−27
Remainder	9.6	92
− Function and value	−4.2	−18
Remainder	8.6	74

(e.g. telephone exchanges) tend to have lower values than others (e.g. office blocks) so the two factors tend to act together. Separate quantification of their effects, even by the most advanced methods of multivariate analysis, is unreliable because the high rates for telephone exchanges may be intrinsic or due to their small size. Only greatly overlapping ranges of values for each building function could separate the effects of the two factors and such overlapping is too infrequent for separate evaluation to be satisfactory.

Another difficulty with representation of an effect on variability by a single number is that the effect of value on unit prices is unlikely to be the same for each building function. It is unlikely to be the same for each element of a building and for each type of work. As buildings with different functions have different proportions of types of work they must be expected to differ in their response to size and therefore to value.

Users of building price data hope for the evaluation of more causes of variability. They want to improve the consistency of their data and their estimates by taking out identified effects; but, unfortunately, as more factors causing variability are identified, the scope for finding important new ones gets less and less. The outlook for reducing variability much more by calculated adjustment is bleak, so it would be wise for those using price data to make adjustments subjectively, but systematically, and to learn to live with variability. The methods developed in this book are intended to make this easier.

3 Selection and adjustment

3.1 Selection of data

When price data from building projects are required, whether for research or for estimating and cost planning, it is to be hoped that there are several projects which could be used as sources. It is argued in this book that it is a mistake to reduce the number of projects to one by narrowing the building description or by any other means. Nevertheless, except for research, extracting data from large numbers of projects could take too long so it is often necessary to reduce the number to between 4 and 10.

In research the population description may be broader and cover more cases than when the analysis is directed towards a particular project. Also, in research, facilities and time are available for handling large quantities of data. All the same, even in research, there are occasions when there are more projects available for analysis than there are resources for analysing them.

In such cases a selection must be made. It is unwise to use judgement in making the selection. This is because the cases from which a selection is required should conform to a carefully worded description of the desired population and any exercise of judgement is liable to imply a change in the description. This may be acceptable but usually is not because at this stage the effect of changing the description can be seen in advance. When using data to test a theory it is easy to convince oneself that it is desirable to omit, say, single-storey buildings when it is seen that the one that is liable to spoil the confirmation of the theory is a single-storey building.

One way of avoiding this danger is to omit the variable to be analysed (usually price level) from the record which describes the building. Then selection can take place without knowledge of the effect on the analysis. The price level could be on the back of the record sheet or in a separate record linked by a reference number. In a computer record it is only necessary to avoid calling up the variable for analysis until the selection process is complete.

Of course this is hardly ever done. In practice the usual preliminary procedure when data have been brought together for the purpose of drawing a conclusion is to check them for 'reasonableness'. If this is merely looking for gross errors such as a missing decimal point or the omission of a large proportion of a project it is useful; but very often it amounts to passing the price variable which is to be the subject of the analysis through a mental sifting process. Those which are reckoned to be surprisingly high or low are simply omitted from further consideration as 'untypical'. Reasons for this are sometimes given which are supposed to justify exclusion; such as, 'Of course that job was expensive – look where it was' or, 'Work was scarce then so it was sure to be cheap'.

It may be true that the characteristics mentioned affected the price but dozens of other influences are ignored. Most projects have several characteristics which could produce a high price but also other characteristics which tend to lower the price. Both sorts can be identified in any project if sought. Trouble is caused when a preconceived belief that the price is high causes only upward-tending characteristics to be sought and vice versa. Whether a price is actually high or low depends on the net effect of the strengths of opposed influences; some upward and some downward. These can seldom be sufficiently accurately quantified for this net effect to be reliably ascribed to particular influences in an individual case.

In this connection it is instructive to carry out experiments on oneself and, if relationships are good enough, on colleagues. A good but perhaps unethical experiment which has been inflicted by the author on unsuspecting colleagues is to present them with the particulars of several projects and to ask them to identify significant influences on the lowest tenders, some of which are stated to be higher and some lower than the estimate. Reasons will be given which indicate that your colleagues carry in their minds two short lists of possible influences: one list to explain high prices and the other low. The same project details are then presented to other colleagues in the same way except that the projects whose prices had been described as higher than the estimate are this time described as lower than the estimate and vice versa. The reasons will be selected from the same short lists as before but this time applied to different projects. The view taken of the influences on a project's price is strongly related to a preconceived idea of the 'proper' price level.

Selection of data by subjective means according to whether they look reasonable or typical is likely to produce answers which have some of the characteristics of self-fulfilling prophecies.

3.2 Descriptions

The correct way of selecting data is in strict accordance with a predetermined description. Great attention should be paid to descriptions to which data are to conform. Poor descriptions leave room for doubt, and therefore subjective choice, about whether a project fits a description or not. They give rise to more serious and undetected errors than arithmetical mistakes.

An example of a careful description might be: 'One- or two-storey local offices built in Yorkshire and Humberside for the Department of Health and Social Services on fixed price contracts based on tenders accepted in 1981.'

Descriptions can be broadened to admit more cases but only if all data conforming to the new broader description are added. A description carries an implication that all members of the described population have been included or, if only a sample is being analysed, that all members of the described population had an equal chance of being selected.

The last remark raises a big subject which causes a great deal of trouble. It is difficult to formulate rules to ensure equal probability of selection which will help in all circumstances. However, in particular cases the main thing is to consider the matter and not to forget it. Whether every member of the described population stood the same chance of being selected when the sample was being drawn is usually obvious once the question has been raised with the person making the selection.

The errors which occur are often due to more than one person being concerned. For instance, when personal approaches are required the gatherer of the data may have more success with, or prefer to deal with, one source rather than another. If this is discovered, even after the event, the error can be corrected by treating the sampling as stratified, provided that the sampling within the source strata can be regarded as random.

3.3 Random sampling

If, after selection according to a description, there is still a need to reduce the number, further reduction should be on a random or pseudo-random basis. There are also other occasions for wanting to select randomly. For example, when the work of others has to be audited it is often important to select work for checking on a random basis so that favouritism can be ruled out and so that a fair picture of the quality of the work can be obtained.

In order of preference, taking into account safety when more than one person is involved, and convenience, the following methods are recommended.

A pseudo-random method of taking a 20% sample would be to take serial numbers ending with either of two specified digits. Ideally the digits should be chosen at random. Different numbers of digits give different percentages. The penultimate digit can be used to break samples down further. Although the intended sampling proportion is only approximately achieved this seldom matters. The method is very convenient and at any stage the question of whether an item should have been included in the sample can be simply answered by inspection of its serial number. It is unsatisfactory only when 'serial' numbers are not in fact serial but have been allocated according to a system. The system might be to give every batch received a new block of numbers the first of which ends in nought.

Another pseudo-random method is to take every nth case to give a $1/n$

sample. Here again any possibility of a systematic ordering of cases must be watched for. It is unlikely but not unknown for cases to be recorded in batches whose size may form a rhythm with the sampling interval. When using this method the starting point for counting should be chosen at random by selecting a random number in the range 1 to n.

A purer method of random selection is to draw cases 'out of a hat'. Such a method is theoretically more random but will not produce more representative samples than the pseudo-random procedures. Statistical theory assumes truly random selection but answers will not be detectably affected by using pseudo-random selection. In fact, where there is any bias introduced by pseudo-random selection it is likely to be in the direction of a very slightly better estimate of the population average.

Another method of random selection which is more convenient than drawing cases 'out of a hat' when they are serially numbered is to number all cases serially and use a random number generator on a calculator or a computer or random number tables. Unless there are exactly 10, 100, 1000, etc., cases there will be some random numbers which will not have been allocated to cases. These numbers can be ignored without affecting the randomness of the sampling.

Although ignoring numbers is safe it is laborious to do so when, for example, there are 102 cases from which a random sample of 10 is required. If a random number generator is used it must be set to produce numbers in the range 0 to 999 so those greater than 102 have to be ignored. In this case it is quicker to use tables than a random number generator because the eye can quickly pass over the unwanted numbers in a table.

However, a random number generator with a range of 0 to 999 can be used for smaller ranges by dividing by a suitable number. Thus dividing by 5 and taking the next lower integer gives random numbers in the range 0 to 199.

A point to watch is that dividing by some numbers leaves a remainder which must not be used. For example dividing 999 by 7 gives random numbers in the range 0 to 142. 143 must be ignored because it is produced from 994 to 999 – only 6 numbers instead of the 7 which give each of the numbers in the range 0 to 142. Using 143 would break the rule for randomness that all numbers must be equally likely.

If a calculator produces numbers in the range 0 to 99, two numbers at a time can be placed side by side to give a four-digit random number in the range 0 to 9999. A three-digit number in the range 0 to 999 can be obtained by putting the first digit of one number beside the other number. Whether the first or second digit is taken and whether it is put to the left or right of the other number does not matter provided, of course, that the chosen procedure is followed consistently.

When zero is not required (and it seldom is) it can be treated as 10 to alter a range of 0 to 9 to one of 1 to 10. Similarly, a range of 0 to 99 can be changed to 1 to 100, and so on.

A computer which produces random numbers in response to a command usually provides them in the range 0 to 1. To convert them to integers ignore the decimal point and use the first one, two or three digits. The same result can be achieved by multiplying by 10, 100 or 1000 and rounding down. Adding 1 will make them start at 1 instead of zero.

3.4 Indices and their use

Among those who deal with building price data there is an almost universal reluctance to use data which are more than two or three years old. The reason usually given is that factors causing price inflation over more than a year or so are much more powerful than any others; therefore differences between prices separated in time are so great that they conceal differences due to other factors.

It is true that adjustments for time have to be very large, but the size of an adjustment is not in itself a reason for doubting its accuracy. The only valid reason for reluctance to adjust for a factor is not that the adjustment is large but that so little is known about its influence on price that the adjustment is unreliable.

With very rare exceptions the only conditions which change rapidly are general economic inflation and market conditions. These can be monitored together and are almost solely responsible for the need for large adjustments for time. Changes in factors for which adjustments are not readily available, such as improvements to construction methods, differential alterations in the balance between plant and labour inputs and between trades, take place slowly and can safely be ignored over periods of 10 years. In any case they are to some extent swept into a simple overall adjustment for time.

Of all the means available for the adjustment of price data the most reliable by far is the adjustment for changes in general price levels over time using a price index. This is done by dividing by the index to bring the price back to the base date of the index (when it was 100) and multiplying by the index at the desired date. It is the author's view that too much importance is usually placed on differences associated with time. Expensively gathered data are too readily discarded as out of date. This is beginning to matter more now that large computer data bases are becoming cheap to use and hungry for data.

Adjustment for date of tender using a published price or cost index is easy and reliable. The reliability derives from the large amount of data upon which the indices are based and the great care with which they are constructed.

Several sorts of index are calculated and each has its appropriate use. The application of those calculated for particular contract conditions, types of work and sectors of the market is obvious. Less obvious is the choice between price and cost indices: price being what the builder charges the client, and cost being, for the purpose of distinction, the cost to the main contractor and subcontractors.

3.4.1 Price indices

Price indices are more generally useful than cost indices and are usually calculated from prices agreed when tenders are accepted. They cannot be based on final accounts because they would be too out of date to be of much use except for historical purposes.

The simplest index is the price per unit area for a particular type of building. To cover more than one building type the price per unit area for each has to be related to a standard price for that type, which may be simply the average price on a given date. This index is valuable because of its simplicity but it is affected by changes in the average quality of buildings. For a particular type of building this can happen suddenly. However, any method has disadvantages and the difficulties with this one are, at least, easy to see. Where design standards are specified for a common building type this sort of index can be very useful. On the other hand, when cost limits are specified it reflects only their movements and the accuracy of estimating.

There are other sorts of index and their methods of calculation differ, but all refer to some sort of standard price for each building contributing to the index.

One method was devised for government work in the United Kingdom but is now used more widely. The largest items in each Bill of Quantities are repriced using a schedule of standard rates as a price book. Any price book could be used if it covered the work well enough. In principle the method expresses the agreed tender price as a percentage of what the price would have been at standard rates. The actual method is more elaborate than this so as to deal with such difficulties as sums of money not allocated to work items or parts of the building and with contracts with price fluctuations clauses.

The number of items selected for repricing need not be many because of the tremendously strong correlations between prices of items in a Bill of Quantities when expressed as ratios to standard prices. The dozen or so largest items making up 25% of the total value will adequately represent the overall price level. The gain from increasing the proportion of repriceable value from 25% to 75% is quite small. In fact, although the effect is only slight, introducing the small items covering the last 25% of value tends to confuse the picture and reduce the reliability of the index.

In practice the formulation of price indices is quite complex, mainly because of efforts to refine them for macro-economic purposes and, at the same time to try to make them represent the price levels of individual projects. For adjustment of building prices these refinements are not required but they do not reduce accuracy and may increase credibility. Apart from refinement there are reasonable grounds for differing opinions on the principles of index construction. There is no single correct method.

An index which uses as quantities the ones actually occurring in the current Bill of Quantities, such as a tender price index which takes the largest items in each Bill, is called a *current-weighted* or *Paasche* type of index. There is also another type which uses a standard list of items and fixes the quantities, taking from the current Bill only the unit prices. This is called a *base-weighted* or *Laspeyre* type of index.

The base-weighted type of index is purer in concept but, with its standard list of items and quantities, has the serious disadvantage that it can omit some of the largest items in a particular Bill. If the index is to be used for historical purposes it will also be distorted by long-term changes in the make-up of buildings. The strong correlations between prices for items reduce both disadvantages but occasionally the standard list of items and quantities, or *standard basket* as it is often called, has to be revised.

Whatever method is used for calculating the individual project indices, they have to be averaged in some way to produce an index covering a period, typically three months. The method of averaging could either give all projects equal weight or allow the larger projects to have their full effect. The latter is described as *weighted by value*. Which is more appropriate depends on the use for which the index is intended.

A value-weighted index is suitable for applying to whole programmes of work including all sizes of project. It is especially appropriate for large-scale financial calculations relating to the building programme of a public body, and for national economic planning when costs have to be expressed at the price level of a different year.

Because its movements are dominated by prices from very large projects a value-weighted index is less suitable for updating prices and estimates for smaller projects. An index for general purposes should be based on the average of project indices regardless of the size of the project. This is an *unweighted* index.

In practice it is unacceptable to have two separate indices for the two purposes. Users of indices expect all price movement information to be contained in a single figure. A sensible compromise would be to weight the indices by the square root or the logarithm of the value but this idea seems to satisfy no one.

If there were enough building projects upon which to base them, the proper solution would be to construct separate indices for different sizes and types of building, but there are never enough data upon which to base more than one overall index. This means that separate movements for different building types cannot be reliably traced. However, this is not of great importance because the relationships between price levels for the various types, sizes and locations of buildings can be assumed to be constant for fairly long periods.

3.4.2 Price factors

The comparatively slow change in the relationships between price levels allows time for enough data to be accumulated to evaluate factors to express

Selection and adjustment 31

the relationships. Where only a coarse split into a few groups is required, for example into three size groups, there will probably be enough data to update the factors annually. For a refined breakdown, perhaps into 30 locations, several years' data may have to be combined.

When more than one quarter's indices are being used they should first be standardized for time by dividing each project index by its appropriate quarterly index. There are theoretical objections to this but they are negligible in practice. Then the standardized indices for each group should be averaged to give a factor for the group.

Indices and factors produced in this way are used to allow a price from one size of building in one location to be used to help estimate for a different size building in another location. It is unlikely that anyone would want to use a price from one building type for another but type factors have to be calculated incidentally to the production of size factors. This is because the effect of size on price level is not the same for all building types. Factors of size must be produced within a type breakdown.

The use of adjustment factors is described in Sections 3.5 and 8.5.

3.4.3 Changes in the mix

Another problem is presented by changes in the mix of projects. An increase in the proportion of a type of building which has a high price level will raise the index even when all price levels actually remain constant. Like the weighting problem this, too, would ideally be dealt with by constructing separate indices but again there are never enough data.

One solution is to decide on a standard mix of building types and reweight the mix which actually occurs to match the standard. This has the same attraction as a standard basket of Bill of Quantities items for calculating individual project indices; but also the same objections apply with even more force. The buildings of a particular type would be given a constant weight in the average however many or few there were. There may sometimes be only one or two, or even none. Agonizing decisions on whether to change the standard mix are frequently required and are often found to have been wrong soon afterwards. Indices based on the old and the new mixes have to be run side by side to provide an adjustment factor to avoid affecting the index level.

A compromise which has been found acceptable is to change the standard mix gradually in accordance with long-term changes while ignoring short-term fluctuations. The technique of exponential smoothing is ideally suited to this and is illustrated by the following example. The example relates to two types of building project but the method is the same for more than two or for the mix of contract types, of regions or any other factors.

In a quarterly index based on both housing and non-housing projects where housing is known to have generally a very different price level from the rest, it would be sensible to stabilize the weights given to the two type groups.

Otherwise movements in the index could be dominated by the proportion of housing projects which happen to have been included in the quarter. If the index is value-weighted then it is the proportion of value accounted for by housing projects which has to be stabilized, but the principles are the same.

First, initial weights have to be chosen. If in the latest year's figures the proportion of housing was 40% and there is no reason to expect a large change in the immediate future, then 0.4 and 0.6 would be suitable initial weights for housing and non-housing respectively. They could however be chosen differently in the light of knowledge of likely changes.

Next, a smoothing constant or, as described here, smoothing weights, must be decided upon. These control the weight to be given to the past in relation to the present. Smoothing weights of 0.8 and 0.2 will be used in this example because a fairly slow response to change is required. A quick response would be provided by smoothing weights of 0.5 and 0.5 and would be desirable if medium-term changes in the ratio of housing to non-housing were to be expected to occur fairly often.

Any pair of smoothing weights can be chosen to produce a desired effect as long as they add up to 1. They can be changed if altered circumstances make it undesirable to allow the past to have the same effect as before.

When an index for a new quarter is to be calculated, the proportion of housing projects in the quarter must be found. If it is 15% it has to be used to modify the initial or past proportion of 40%. In other words, the initial weights of 0.4 and 0.6 are to be combined with the current quarter's of 0.15 and 0.85. They are combined using the smoothing weights of 0.8 and 0.2 as follows.

	Old weights	Current quarter	Old × 0.8	Current × 0.2	Combined weights for current quarter
Housing	0.40	0.15	0.32	0.03	0.35
Non-housing	0.60	0.85	0.48	0.17	0.65

If in the next quarter the proportion of housing is 25%, the new weights are calculated as follows.

	Old weights	Current quarter	Old × 0.8	Current × 0.2	Combined weights for current quarter
Housing	0.35	0.25	0.28	0.05	0.33
Non-housing	0.65	0.75	0.52	0.15	0.67

Where, as in the above example, there are only two groups to be smoothed it is not necessary, except as a check, to calculate the weight for the second because it can be obtained from the first by subtraction from 1. It is calculated above to show how the two relate to each other and because it makes a useful arithmetical check. In some cases where rounding has taken place, slight adjustments to the combined weights will be necessary because they must add up to 1 when they are applied.

Smoothing the weights is not the same as smoothing the index itself. The weights are smoothed to reduce their effect on total variability. However, there is a slight incidental tendency to smooth the index and make genuine changes in price level a little plainer.

Applying a smoothing technique to the index itself rather than to the weights is a crude way of reducing erratic movements which must, at the same time, delay the effect of a genuine change in the trend of the index. There would be less delay the more the smoothing weights tended towards the current quarter but also, of course, less smoothing effect. Whenever smoothing weights are being chosen it is instructive to try their effect by applying them to the past history of the series. Each quarter is then taken in turn as the current quarter. Different smoothing weights can be tried until a pair is found which produces the desired effect.

When a series of numbers, such as the weights in the above example or the quarterly index itself, is subjected to exponential smoothing the smoothed figures are inevitably older than the current quarter. Their average age depends upon the smoothing weights and can be determined by the following formula.

$$\text{Age (measured in index intervals, e.g. quarters)} = \frac{\text{weight for past}}{\text{weight for present}}$$

In the example the weight on the old data was 0.8 and on the new was 0.2, so the average age was $0.8/0.2 = 4$ quarters $= 1$ year.

When the numbers being smoothed have no definite long-term trend there is no need to adjust for the age of the smoothed figures. On the other hand if, as is usually the case with a price index, there is a tendency in one direction, a trend correction should be applied. The trend correction, to be added to the smoothed index, is the average age as calculated above, multiplied by the amount by which the index is changing in one index interval. This amount is simply the difference between the last two smoothed indices, taking the sign of the difference into account, recorded as another series and smoothed in the same way as the index itself.

One use of an exponentially smoothed index is as an aid to forecasting. Not too much should be expected of the technique but, as far as it goes, it is a good way of revealing underlying trends to improve understanding of index movements. It is discussed in Chapter 10.

3.4.4 Logarithmic plotting

If a straight-line method of projection is to be used it is theoretically better to work with the logarithms of the index. This is also a good way of plotting the index, because a constant inflation rate then comes out as a straight line instead of a gradually steepening curve. Logarithmic graph paper can be used to plot the indices directly, or the logs of the indices can be plotted on ordinary graph paper. There are many possible scales and combinations of scales for log graph paper so it is quite likely that the right one is not available when needed; therefore it is useful to know how to use ordinary graph paper.

The following example shows how to plot on a logarithmic scale using ordinary graph paper. The base of the logarithms does not matter. Calculating machines and computers give logs to base e more readily then the more familiar logs to base 10, requiring only one keystroke on a calculator instead of two. Logarithms to base e are just as easy to use and will be used in the example.

An index inflating at 10% per time period beginning at 110 would run as follows.

Time period	1	2	3	4	5	6
Index	110	121	133	146	161	177
$\log_e$ (index)	4.70	4.80	4.89	4.98	5.08	5.18

If the graph in Fig. 3.1 is to be published, the log scale on the left should be

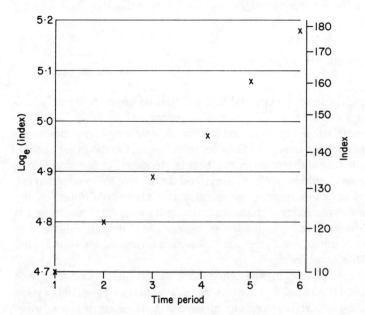

Fig. 3.1 Logarithmic plotting of an index inflating at 10% per year.

replaced by the scale on the right. In that case the horizontal lines would be drawn so as to relate to that scale.

Because, to any base, the logarithm of 1 is 0, it is unusual to start a scale below this. There is no reason why logarithms of numbers less than 1 should not be plotted as negatives but the scale can never go down to the log of zero because this is minus infinity on any scale.

3.4.5 Cost indices

Tender price indices are an attempt to represent the level of prices agreed between clients and contractors. They move according to changes in contractors' costs but also with the market as contractors, for example, cut their prices when competing for scarce work or increase them when taking advantage of a glut of work. They also vary, over a long time, with changes in productivity.

When a contractor wishes to use data from past work as a guide to what is likely to happen on a new job, he can use operational data and direct measures such as quantities of materials, man-hours and plant-hours. He can make adjustments to these for changes in productivity and methods. Often, though, he will want to work with costs and wish to update them directly using cost indices.

For such purposes market variation must be excluded. The contractor will want to add his own allowance for this at the time of tendering so he will not want it already included in a tender price index. Ideally he will want an index of movement of his costs which takes productivity changes into account but which excludes market variation.

A very large contractor can construct his own index of costs of his own work using one of the methods by which price indices are calculated, but instead of using the prices he charges he uses his calculation of his costs.

If the contractor is prepared to assume that productivity does not change he can construct an index from the average amount of materials, plant and labour per unit floor area used in one or more common types of building in each time period, for example a quarter, and apply current costs to them.

If he wants to give each project of a given type equal weight he will divide by floor area individually for each project. Otherwise he will add the quantities for all projects together and divide by total floor area to give weight in proportion to area. Movements of such an index will be dominated by the largest projects. As in the calculation of tender price indices, a compromise is possible by using the square root of floor area as the weight.

The contractor may not wish to use his own cost records and prefer to use published materials, labour and plant cost indices. These can be applied to his own quantities to show how average contractors' costs are moving for the sort of work he is doing.

To do this he simply adds up the total quantities of the main resources used in, say, a year and applies current cost indices to these quantities each month or quarter.

A very large contractor may have enough data to keep separate indices for each major category of work or for each type of building. This is preferable because the relative movements in indices help to guide the types of work to be most actively sought. This is important when, for instance, labour costs are rising faster than materials costs.

Different methods of calculation must obviously give different results but indices are only meant to measure changes in cost or price levels. It is the ratio between the indices for different times that is the measure of change so, provided that the same method of calculation is used throughout, comparisons from one time to another will be valid: that is, valid in their own terms. They can only reflect what goes into them and are no more than an agreed method of representation. There is no single right method and the one used should always be borne in mind when using an index. If, when making a single comparison of costs without an index, it would seem inappropriate to use the method underlying the index, then the index is inappropriate for such comparisons. This obvious statement is the answer to many questions asked about the use of indices.

There are times when an indexing method has to be changed to suit changing circumstances, availability of data or because a great improvement has been devised. The new method will affect the level of the index so the new index level has to be made to match the old. For a few time periods both methods must be run side by side and the ratios of the results averaged. This average ratio has to be applied permanently to all indices calculated by the new method.

3.4.6 National cost indices

For macro-economic purposes national indices are constructed. National statistics of total quantities of resources used serve as weights to apply to published resource costs. Making assumptions about productivity changes allows the calculation of macro-economic cost indices.

These indices can be used not only by economists but also by those who need indices which exclude market variation but who cannot construct their own. Cost control during design can reasonably have as its objective the minimization of contractors' costs or, rather, the balancing of them against design considerations. Market variation is outside the designer's control so he may prefer to measure contractors' costs. When it is necessary to update earlier measurements of costs a cost index, rather than a price index, is appropriate.

Ideally the proportions of materials, labour and plant used in the construction of the chosen index should be appropriate to the type of work which is the subject of the cost control, but this is seldom possible.

3.4.7 Formula methods

There are cost indices which are constructed for special purposes to provide an agreed basis for an inexpensive assessment of cost changes. They can be tailored to suit the application and, provided that their construction is thoroughly understood by the users, can command sufficient confidence to replace far more laborious and expensive methods of calculation.

This is the purpose of the indices which are constructed to feed formulae designed to measure changes in the costs of work sections in Bills of Quantities. These allow the adjustment of sums paid to contractors to compensate for changes in their costs since the base date to which the original contract related.

The formulae and indices have become quite elaborate but they hold no special interest from the statistical point of view. The degree of refinement is decided mainly by what is required to satisfy the doubts of all parties to the contract.

Formula methods provide only rough justice. Inevitably some quite large over- or under-payments will occasionally result from any formula method whose complexity is much less than the full and exact calculation of changes in costs. At such times confidence is likely to be shaken and there is a tendency gradually to increase the complexity of the indices to satisfy these doubts. Frequent revisions aimed at simplification can combat this deficiency.

3.4.8 Ratio of price to cost

Both sorts of index, price and cost, are used by economists. One interesting property is the indication of movements in the construction industry's profitability and productivity given by the ratio of a price index to a cost index.

The effects of changes in profitability and productivity cannot be separately assessed in this way but, except over very long periods, it is reasonable to assume no change in productivity so movements in the ratio can be ascribed to changes in profitability. Just as with indices themselves, changes in the ratio are entirely relative to each other. Absolute values mean nothing. A movement in the ratio of indices from 1.2 to 0.9 does not necessarily mean that the industry has started to run at a loss. It just means that the true ratio of price to cost is now only three quarters of what it was. If it had been 1.6 it would now be 1.2; if it had been 1.4 it would now be 1.05. If a crude definition of profit is used, this could mean that it has fallen from 60% to 20% or from 40% to 5%; if it had been 20% it has fallen to -10%.

3.5 Adjustment of data

3.5.1 Adjustment factors

The calculation of adjustment factors using tender price indices for building projects has already been described in Section 3.4.2. It amounts to classifying the projects according to the groups for which factors are required and averaging the indices in each group. In practice, however, there are seldom enough data in many groups to provide a reliable average. Adjustment factors need to be fairly accurate so 20 or 30 projects are needed for each average. There may be enough schools or offices to allow a coarse breakdown by size and, possibly, type of construction but it is seldom possible to go further.

A refined method of analysis which calculates adjustment factors for all the effects at once is multiple regression analysis, described in Chapter 8. The technique makes best use of the available data but its assumptions must be carefully considered in each application. No refinement of analysis can do much to make up for lack of relevant data. Often the extra effort put into analysis would have been better employed in gathering more data for simpler analysis.

One cause of lack of data is the need to subdivide in order to isolate types of building. One way out of the difficulty is to change to another method of measuring the effects in question.

A suitable method which does not require the analysis of data is to use estimating techniques to arrive at a cost for a design. If the design is then changed in carefully chosen ways the effect of, for instance, height, shape, type of roof or type of cladding can be calculated by repeating the estimating process.

3.5.2 Subjective adjustment

If adjustment factors cannot be calculated they may have to be guessed, remembering that failing to make an adjustment is, in fact, an assumption that no adjustment is required. Although it is dangerous to be subjective in the selection of data, it is less hazardous to be so in framing rules for the adjustment of data to compensate for influences which are believed to have affected them.

The total effect of adjustments is probably less than expected. Some are up and some down. Their effect on variability is unlikely to be great, so even if they are ill-founded little harm will be done. If insufficient analysed data are available it is probably better to formulate adjustment rules subjectively than to ignore influences on the data which are known to affect them but which have not been properly measured. Ignoring an unquantifiable effect is not playing safe; it is making an assumption that is known to be wrong. It is assuming that the effect is zero.

This encouragement to guess should be accompanied by the warning that subjective ideas of the effects of influences on price tend to exaggerate them. Therefore subjective adjustments should be moderate. The most important proviso in recommending subjectivity is that the adjustments must be made, in one respect, blindly. That means that they must take no account of whether the price seems high or low. This is possible only if the adjustments are decided upon without knowledge of the price. Variables for analysis, such as price or energy cost, are best recorded separately from the descriptive factors upon which adjustments may be based. This is easiest when the data are recorded in a computer data base.

3.5.3 The importance of adjustment

Adjustment of price data, especially if it is subjective, is contrary to normal practice. Using judgement to reduce the number of projects, or other sources of price data, to the 'ideal' one is regarded as good practice but adjustment so as to retain several projects as the bases of an estimate is not.

A single item of data can never be relied on because it is only one member of a wide-spread distribution of such items, any one of which could have arisen by chance. The only way to increase reliability markedly is to average several of such items.

The common preference for exclusion rather than adjustment may be because exclusion is usually unrecorded whereas adjustments are normally declared. Exclusions are thus unlikely to be a subject of dispute whereas no two people would agree on the proper size for an adjustment.

Data are commonly categorized as either good or worthless. In fact the differences in reliability of data are less than is generally believed. Most of those condemned as useless could, with adjustment, be almost as valuable as the rest.

Adjustment factors are important to the organization of a data base, especially a computer data base. The computer retrieval program can adjust the data according to stored factors and enable a wider variety of projects to be considered and thus a greater number.

In order to make maximum use of scarce data it may be necessary to use methods of adjustment which would shock a purist. If methods to be described shortly are used, the spirit of objectivity must be kept in mind even though subjective methods are being used. The danger of introducing bias must be watched for.

If honest attempts are made to avoid bias such methods cannot be condemned because they are imperfect. Their derivation is easy to understand and their use tends to move results in the proper direction more often than not. This is sufficient justification for using them.

3.6 Blending data

When objective data are too few to rely on they can usually be augmented by other data which have been adjusted subjectively. If this is not possible it may be necessary to take the final step away from objectivity and use a subjective opinion of the correct figure. It is better to use such a 'guesstimate' deliberately than to allow a subjective idea of what the figure ought to be to influence either the selection of data or the adjustment of selected data. Such a guessed figure can be included in an analysis alongside existing data.

There is a branch of statistical theory, little used until comparatively recently, known as Bayesian methods which deals with the blending of new data with existing information, but its techniques are not necessary for the analysis of building price data. The data are too variable and their analysis need not be refined.

Blending of data from different sources can be dealt with quite simply. The crude method described here is good enough and far better than the usual choice of a single source and the rejection of others.

Treatment of new data should not be on such an accept-or-reject basis. They should not replace existing knowledge completely neither should they be completely rejected. The information from the new data should be blended with existing information so as to modify it; the extent of the modification depending on the relative degrees of confidence which can be placed on the old and new information.

To be systematic, whenever numerical data are recorded for future use an assessment of the confidence which can be placed in them should be recorded with them.

3.6.1 Weighting data

The same degree of confidence cannot be placed in all data. For one thing the sources may differ and some of them must be considered to be more reliable than others. Again some data will have been adjusted by less reliable factors than others.

The degree of confidence can seldom be measured objectively but nevertheless ought to be taken into account when finding some sort of average to represent all the data.

Statistical methods are not appropriate to measurement of these different degrees of confidence but can be used after the subjective evaluations have been made.

The simplest method is to assign weights to the data to represent the degree of confidence and to record the weight as part of the description of each item of data. This can be done on any basis of subjective judgement which seems valid to the user of the data. The vital condition is that consistent criteria should be used. For small quantities of data, weights of 1, 2 or 3 would usually suffice but for large quantities it may be considered reasonable to use a 5-point rating scale.

Data to which weights have been assigned can be averaged by calculating the weighted arithmetic mean. This is done by multiplying each item by its weight, summing the products and dividing this total by the sum of the weights.

To find a weighted median the data are put in ascending order but, instead of simply counting through to half-way as for an unweighted median, the weights are accumulated until half the total weight is reached. The rest of the calculation is explained in Section 6.3.8.

Instead of a single assessment of a weight, a refinement is to obtain it by totalling several separately allocated ratings. The ratings refer to a set of criteria and the scale for each is chosen to reflect the importance of each criterion.

The criteria should be simple and their number few. For example, three criteria could be as follows.

	Score
Source	(max. 10)
Own work	10
Colleague	7
Cost information service	5
Construction press	4
Background detail	
(to assist adjustment)	(max. 3)
Great	3
Average	2
Little	0
Reliability of adjustments	(max. 5)
Good	5
Average	3
Poor	2

3.6.2 Measuring the variability of weighted data

It is possible to compute a weighted standard deviation but its meaning is unclear and the effort is unjustified. It is sufficient to calculate a standard deviation ignoring the weights because in construction cost data the weights are unlikely to be precise so they will not reduce variability by much. An unweighted standard deviation will be a conservative overestimate of the weighted standard deviation.

If a weighted median is being calculated then a weighted interquartile range can be obtained at the same time.

3.7 Data banks

Selecting and organizing data can be laborious unless they are stored in a readily accessible form. The concept of a bank of data, into which they can be put and from which they can be withdrawn, requires some form of 'loose-leaf' storage.

The longest-lived method of storage is the card index. This has been in use

42 Statistical Methods for Building Price Data

so widely for so long that it need not be described here. Worth describing are the hand-punched versions in which holes are punched in the cards to allow selective retrieval.

3.7.1 Edge-punched cards

The commonest type of hand-punched card has pre-punched holes round the edge which can be converted to V-shapes by means of a ticket-punch. Each building in the data bank has a card of its own in which each hole represents a characteristic, a number or a code representing a range of numbers. Punching a 'V' indicates that the characteristic, number or code applies to the building.

Inserting a needle through a hole in a pack of the cards, lifting and shaking it causes the V-punched cards to be left behind. These are the ones selected. Inserting the needle through other holes in those selected narrows their description further.

Ingenious, economical ways of recording numbers have been devised for these cards. Two representations are shown in Fig. 3.2. The number is 85.

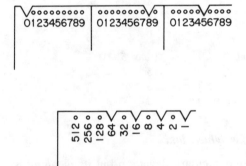

Fig. 3.2 Two ways of showing the number 85 on edge-punched cards.

The upper representation in Fig. 3.2 is normally used unless space economy is vital. The lower representation is the binary system and is now widely familiar because it is commonly taught in schools. It is seldom used, except by those concerned with computer logic, because although it is the basis of computer arithmetic all contact with the user is in decimal form.

In practice it is seldom necessary to represent numbers greater than 9. They are not normally required for purposes of selection, except when coded in broad-banded form such as:

        1 for 0 to 99
        2 for 100 to 499
        3 for 500 to 999

and so on. The actual numbers can be written on the card for calculation purposes after selection.

Unless cards can be specially printed with each hole labelled, it is helpful to make a labelled master card on which every hole has been punched. This can be placed over each blank card when punching and over the pack when selecting.

When punching cards care must be taken to make all the punches deep enough to make a smooth V. Inverted key-hole shapes like the following will cause retention on the needle of punched cards.

3.7.2 Captive punched cards

A disadvantage of edge-punched cards is that it is difficult to preserve the cards in their original sequence. This is a nuisance if the cards are to be updated or used as sources of individual reference like a conventional card index.

If a selection method is required but the cards must be kept in sequence, it may be worth considering a more expensive captive version which is also easier and quicker to use and more reliable.

In one captive-card method the cards are held in a frame by rods through vertical slots on each side of the cards so that when the frame is rotated through 180° the cards will fall about an inch or less.

At each end of the frame is a fixed board bearing a matrix of holes, usually 10 × 10, through which rods can be passed. The rods also pass through matching pre-punched holes in the cards which restrain them from falling when the frame is rotated.

When a hole is hand-punched with a slot joining it to the hole below, the card will not be restrained by a rod passing through the hole.

To select cards which have been punched in a particular position, a rod is passed through the appropriate hole and the frame rotated. Cards with the relevant hole punched will fall, if encouraged to do so by shaking or banging the frame, as far as the slots will allow. Inserting another rod through special holes near what is now the top of the frame will prevent the selected cards falling back to their original position when the frame is rotated to its upright position again.

Any number of rods can be inserted before the frame is turned upside-down and only cards punched in all the positions at which rods have been inserted will fall when the frame is inverted.

44 Statistical Methods for Building Price Data

The tops of the cards should bear identifying particulars and other data which will be needed when cards have been selected. These can be read from them when they project above the others. Examples are name and location of project, floor area, cost per unit area and total cost of building. The backs of the projecting parts can be used as well as the fronts.

Instead of a series of selections, as is required to sort on several variables with edge-punched cards, all the necessary rods can be inserted at the same time. If too few cards are selected a rod can be withdrawn without returning the frame to its upright position.

Rods must not be inserted while some cards are projecting because they will be passing through the wrong holes in these cards.

Amendments to cards can be made with adhesive tape but care is needed to prevent the cards becoming even slightly sticky.

3.7.3   Optical coincidence cards

If the sole object of the card index is to retrieve data by selection and, unlike the usual card index, it is not required for individual reference, the optical coincidence or feature card system can be an excellent method for data retrieval. It can have a surprisingly large capacity but, if used to the full, punching becomes laborious.

Instead of punching one card per item in the data bank, each card represents what in the other systems are represented by holes. These can be described as features. For instance, there would be a card for 'flat roof', another for 'steel framed', another for 'load-bearing brick', another for 'office accommodation', and so on.

Each feature card is pre-printed with a matrix of serial numbers which represent the buildings. These are the positions for punching holes. It is possible to buy cards with as many as 10 000 serial numbers but this is far more than required for a building price data bank and sufficiently accurate punching is very difficult.

Taking one building at a time, the cards for features which apply to it are removed from the whole set and made into a pack. A hole is drilled through them in the position of the serial number of the building and the card returned to the main pack. Hand punches are available but they can only deal with one card at a time. They are useful for amendments but for speedier work drilling machines are used by specialist firms who supply the cards and provide a complete service.

Unlike the other systems, these cards do not suffer from problems of sticking to each other because they are not required to slide. To select data the cards relating to the required features are taken from the pack and put together. When they are held up to a light, holes which coincide in all the cards are easily identified. Their serial numbers relate to the required buildings. To change the description of the buildings selected, cards can be added or withdrawn. The effect on the number of buildings selected can be quickly seen.

The system is elegant and, once set up, very effective and quick to use. However, accuracy of drilling can be a problem and amending cards is not very satisfactory because tiny patches come off and sealing compounds drop out. Repunching single cards is laborious with a hand punch.

Card index methods are cumbersome in comparison with computer methods but they have, for some users, overriding advantages. They are cheap, portable and immediately accessible. Until cheap computer facilities are always beside the desk and on call within seconds, the humble card index and its punched versions will have their uses in the retrieval of building price data. Their value has been increased by the improvement and cheapness of desk calculators with features including automatic mean and standard deviation.

3.7.4 Machine-operated punched cards

The Hollerith card, which is now a common means of input to computers, was designed for direct sorting and tabulating by machines. These lack the portability and accessibility of hand operated punched cards and the speed and flexibility of computers. They are hardly used now.

3.7.5 Computer methods

Large quantities of data can now be cheaply stored on magnetic disc and rapidly retrieved so the disadvantages of serial storage on tape have been overcome. The value of computer methods lies not only in their rapid data retrieval but in the ease of data input, amendment and output, and in their power of analysis once data have been retrieved.

The value of computer methods depends on the programs which are available – the software. Ready-made general purpose software may be unsuitable or may require too much storage capacity. Specially written software is prohibitively expensive but can sometimes be written by the user who enjoys programming and does not apply the usual costs to the time spent.

Data retrieval library routines or packages are available for most computers but the handling of the data once retrieved is usually poor unless a separate statistical package is used. With a large computer system, such as those available on time-sharing networks, this is sometimes straightforward, but most users would prefer to use a microcomputer. For these the interface between the process of selecting data according to a description and of statistical analysis is at present unsatisfactory. Also the way missing data are handled and the capacity of the statistical analysis package may be inadequate.

The interface problem is less for the larger computers because they have better means of temporary storage of data while loading the statistical package. With systems such as these data can be loaded, samples selected and analysed, other samples selected, and so on, during such a simple dialogue that the operator's mind can be concentrated on the problems of analysis. However, for most computer users the method of analysis must be provisionally decided beforehand and carried out in programmed stages. Plenty of time must be allowed between stages for rethinking in the light of earlier results. At least, such a way of working greatly reduces the temptation present in a simple interactive method to analyse thoughtlessly.

The user should not be content until he has, first, easy access to a large and frequently updated data bank so that he can quickly select the data he needs and, second, a cheap and flexible means of analysis. This seems to indicate a link from a central computer to his own office so that he can select data which are then copied to his own microcomputer where he has suitable software for analysis.

Computer statistical packages make available advanced statistical methods at little, if any, more expense than elementary methods. This provides great analytical power but also brings the danger of using techniques instead of thought. Many elementary errors are caused by giving more attention to the elegance of a technique than to its meaning and relevance for solving the problem in hand.

Those concerned with building need less warning than most of the dangers of using advanced techniques which they do not fully understand.

3.7.6 Data structure

Structuring the data in a building price data bank is usually fairly straightforward because there is seldom much choice of arrangements. However, there are a few points of difficulty.

Normally each case will be a building and its data will be a mixture of descriptive and quantitative variables. Data are often unavailable so careful consideration must be given to the representation and the handling of these missing data. The possibility that missing values for quantitative variables will be confused with zeros must be guarded against. For descriptive variables there is the likelihood that the 'not knowns' are different from the others in ways connected with the difficulty in collecting data for them. They must be coded in such a way that instead of being omitted from analyses they can be treated as a special category and studied more carefully rather than less.

A common problem with building price data is that many cases fall into more than one category of some of the descriptive variables. For example, type of construction is likely to be a descriptive variable and many buildings are of mixed construction.

Such complexities are inseparable from building price data. They must be lived with and not concealed. There are no entirely satisfactory ways of dealing with them and each person must develop his own methods which suit his needs and which can be applied consistently. They must not oversimplify merely to produce a tidy solution. For instance, the mixed construction problem could be solved by always deciding which is the dominant type for each building or by ranking the types and always using the types which, of those relating to the building, is highest on the list. Both would be tidy solutions but may be found unsatisfactory when analyses were made. Some mixtures are common enough to demand categories of their own.

Rules for coding can be formulated to suit the type of analyses which will be made and these rules must be borne in mind at the time of analysis and, especially, while conclusions are being drawn. However, some judgement has to be used in applying the rules. Because the interpretation of results is intimately bound up with judgements made at the time of classification the person concerned with one must be involved with the other.

One way of minimizing the chance of misunderstanding between the coder and the analyser is for the coder to use such a detailed code that decisions on grouping are left to the person carrying out the analysis. Such code lists can be very long and it is impossible to eliminate all need for judgement during coding. Nevertheless, if the time can be given to the coding and a ready-made list is available, it is the best way. A well constructed code is easy to use and allows useful groups to be formed during analysis. The copy used by the analyser should be freely annotated with, for example, reminders where items elsewhere in the code should be considered.

An example of an accepted code is the CI/SfB classification of building functions. Even if it is not used in its entirety it would be wise to ensure that any home-made code harmonizes with it by being either an expansion of part of it or composed of groups from it. It is common to omit the last digit except in an area of special interest where even the full code may not be detailed enough and an extra digit beyond the one in the code may be needed.

Use of a standard code or one harmonized with it helps comparison with other analyses and may permit easy augmentation of the data bank with data from elsewhere.

Building value is sometimes used as a descriptive variable to indicate size. Recording in full and grouping into value bands at the time of analysis is safe enough but classifying value into bands at the time of coding is not. This is because values change with inflation and the sizes of building included in a category change. Floor area is a better basis for classification unless value is adjusted to a constant base year before being classified. Generally speaking, grouping should not be done at recording or coding stage. Numbers are best copied from the original record into the data bank without grouping. Grouping can be part of the analysis.

Values can be compared from one time to another if they are adjusted to a common date using a price index. This can be done if the currently quarterly or, ideally, interpolated monthly price index is recorded as part of the data for each building so that values and costs per unit area can be deflated during the analysis to a common date chosen at the time of analysis. However, many find it more convenient to adjust to a base year at the time of coding even if the base year needs to be changed at the time of analysis. Multiplying all by a common factor is easy and staff time may be more valuable during analysis than during coding.

It is advisable to leave plenty of space in the data bank not only for more data but also for more variables. One should be prepared to make retrospective changes after some experience of using the data bank. It must always be made easy to inspect the original data so that a coding policy can be changed retrospectively and doubts about what had been done investigated.

Finally, to repeat warnings given earlier, data should not be excluded from the data bank on the grounds of the price being unlikely. If the case would have been acceptable had the price been more in line with others it should be put into the data bank. Some time later, when more data have been gathered, the case may appear simply at one end of a distribution and be able to play its part in giving a full picture. Methods are given later for dealing with such cases in an analysis.

4 Patterns in prices

4.1 Distributions

When several numbers are being appraised together, the way in which they lie in relation to one another is called their *distribution*. They may tend to concentrate around their average, stretch out more one way than the other or, very much less often, form two clusters. Examples of these types are as follows:

(a) Symmetrical   12, 16, 19, 20, 21, 21, 22, 23, 26, 30
(b) Stretched upward   12, 14, 15, 16, 16, 17, 20, 26, 32, 46
(c) Stretched downward   5, 24, 31, 36, 40, 40, 41, 42, 45, 50
(d) Two clusters   10, 15, 17, 17, 20, 34, 38, 40, 41, 46

Distributions can be plotted on graph paper by making the horizontal scale equal value bands of the variable which the numbers represent, e.g. cost per unit area, and the vertical scale the frequency with which they occur. This is called a *histogram*. Figure 4.1 is a histogram of the data set (b) above.

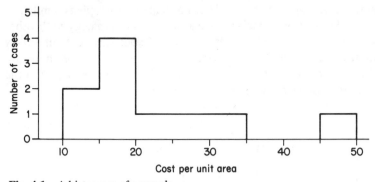

Fig. 4.1   A histogram of a sample.

4.2 Skewness

Variability has been discussed in previous chapters only as it affects the breadth of a distribution: its dispersion. This chapter is concerned with the shape of the distribution.

Almost certainly the distribution of a population of prices will have a single hump. Samples from it may, by chance, be double humped but if the whole population is found to be so then it is probably a mixture of two populations.

Even though a population distribution is single humped and tailing off to zero on each side, the rates at which the two sides approach zero may well differ. This effect is called *skewness*. As there is usually more freedom for costs to be very high than very low, if only because they cannot be less than zero, skewness is usually to the right.

A typical population distribution would look like Fig. 4.2. Because of the large number of cases the curve is smooth.

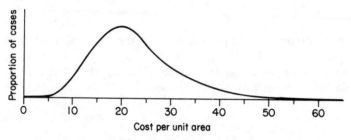

Fig. 4.2 A population frequency distribution.

The sample whose histogram is in Fig. 4.1 could be a random sample of 10 cases from this population. In fact, the sample is even more skewed than the population. However, another random sample from the same population could have a different shape, such as that in Fig. 4.3.

Because skewness is not always the same in a random sample as it was in the population, the best guide to the underlying shape of a sample is familiarity with the population from which the data have come. Analogy with other populations in the same field is also helpful. Experience is built up by examination of data whenever the opportunity occurs. Where there are enough data, more than about 20 items, it should be clear from a plot of the frequency distribution whether it is skewed or not.

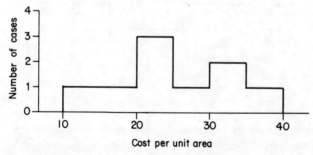

Fig. 4.3 A possible random sample from the population in Fig. 4.2

Although plotting improves understanding, it is not always necessary because a guide to skewness is provided by an examination of the relative positions of certain parameters. This can help even when there are too few data for plotting to be a useful guide. Deductions are bound to be shaky when based on fewer than about eight cases.

The parameters of most use are the arithmetic mean, the median (the middle value in the distribution), quantiles and the minimum and maximum. Which quantiles are most useful depends on the number of cases. It is best to use quartiles unless there are more than 15 cases, when deciles can be used as well.

In the following example the judgement of the likely shape of the parent population is rather difficult. The data have been arranged in ascending order. This is always good practice and eases the task of calculating quantiles, even for a few cases. There are also spaces to mark the class intervals which would be used in a frequency distribution. This would hardly be worth plotting for so few data.

$$8, \quad 11, 15, 16, 18, \quad 23, 27, 29, \quad 35, 36$$

Arithmetic mean	21.8
Maximum	36
Upper quartile	29
Median	$20\frac{1}{2}$
Lower quartile	15
Minimum	8

The difficulty with this example lies in the smallness of the sample and the absence of an isolated high figure commonly found in skew distributions. The maximum is not much further above the mean or median than the minimum is below. Also the mean is only a little higher than the median and, in fact, has five items greater than it and five smaller.

However, the best indication of skewness, after noticing that the arithmetic mean is greater than the median, is the positions of the quartiles in relation to the median. The upper quartile in this case is further above the median than the lower quartile is below it.

The maximum and minimum are erratic indicators so, as both the relationship between the arithmetic mean and the median and the positions of the quartiles support the conclusion that the sample is skewed to the right, it is reasonable to believe that it comes from a similarly skewed population. Of course, this conclusion could be wrong. Such a sample could have come from a symmetrical distribution. Full weight must be given to what was believed about the population distribution before the sample was examined.

Skewness to the left is rare but it can be produced where there is a very strong limitation on the upper level to which prices may rise. This can result from the application of cost limits. In theory this could be so marked that the effect is not to produce skewness but to 'pile up' cases at the right of an otherwise symmetrical or right-skewed distribution.

In practice most effects on building price data operate only coarsely. Cost limits may operate on only part of the building or abnormal circumstances may permit the limit to be exceeded. There may also be other causes of blurring of the effect. The resulting distribution can usually be approximated by a smooth symmetrical or slightly skewed distribution.

4.3 Transformations

When a distribution is known to be skewed it simplifies assessments and comparisons to transform the data to make its distribution approximately symmetrical. Transformation consists of carrying out a mathematical operation such as taking the square root of each number. All calculations can then be carried out on the transformed data and all conclusions based on analysis of them.

Without transformation, analysis of distributions which are skewed to the right puts too much emphasis on high values for some statistical procedures to be strictly valid. Even a simple comparison of two arithmetic means by inspection is more affected by differences in the higher values than in the lower values. This is because random differences in the higher values tend to be greater than in the lower values.

Another reason for wanting to transform data is the desire to change the relationship between two variables from a curved line to a straight one so as to simplify analysis. The principles of transformation are the same.

When transformation has been decided upon the method chosen depends on the degree of skewness.

4.4 Logarithmic transformation

If the ratio of an upper quantile to the median is about the same as the ratio of the median to the corresponding lower quantile, this indicates that the skewness would be removed by a logarithmic transformation. This is done by taking logs of all the data.

A log facility is available on many calculators so it should seldom be necessary to use log tables. Numbers less than one have negative logs so, if this is inconvenient, data which include numbers less than one should be scaled up by a factor such as 10, 100, or 1000 to make the lowest number greater than one. This merely adds a constant to the logs so the effect is easily allowed for in the analysis. Zeros, whose logs are minus infinity, may usually be replaced by a small fraction such as 0.1. In some analyses zero may be interpreted as 'not applicable', in which case they may be omitted or treated separately.

The reason that taking logs removes skewness is that it turns division into subtraction so the two equal ratios would become equal differences. Quantiles symmetrical about the median indicate a symmetrical distribution. Thus a logarithmic transformation has converted a distribution which is skewed to the right into a symmetrical one.

Returning to the ten items of data which were analysed and discussed in the example in Section 4.2, the ratio of the upper quartile to the median is the same as the ratio of the median to the lower quartile, namely 1.4. However, the ratio of the maximum to the median (1.8) is much less than the ratio of the median to the minimum (2.6). Although the indication from the extremes is less important than that from the quartiles, it cannot be entirely ignored when the ratios are so different.

The missing ingredient for a confident judgement is experience of how such data would be expected to be distributed. If a skewed distribution, corrected by logarithms, were expected then the evidence from the sample is enough to confirm the appropriateness of the transformation. If the expectation were for a symmetrical distribution it would be best to leave the data untransformed because it is better not to transform if in serious doubt. If there were no basis for an expectation then a logarithmic transformation would be justified in a refined analysis but not for a simple analysis.

If an analysis has been made with transformed data, internal and external comparisons can be made without referring to the original figures. Such conclusions should be based on the transformed data. When figures are quoted, however, they should be in the original units. Quantiles can be reconverted by taking antilogs but the antilog of the arithmetic mean is the geometric mean of the untransformed data. Rather than retransform it is better to go back to the original data to calculate means and standard deviations.

It should not be thought that a distribution that requires a logarithmic transformation to make it symmetrical is in any way 'unnatural'. There is nothing unnatural in differences or changes being related to size as in a distribution which is rendered symmetrical by logarithmic transformation. There are many fields in which larger values are liable to change more than smaller values and where the amount of change is proportional to size. For example, when costs per unit area are assembled for a type of building, those at the cheapest end of the distribution may have little scope for difference whereas the dearest designs may be designed to high specifications with many ways of differing from each other.

It is not only distributions which require transformation. As already described in Chapter 3, they can be used to change the relationship between two variables from a curved line to a straight one. The variables could be price and time. If prices subject to inflation at a constant rate per cent are plotted on a plain scale they rise in an ever steepening curve, whereas on a log scale they follow a straight line. This was illustrated in Fig. 3.1.

4.5 Square root transformation

For moderate degrees of skewness the less powerful square root transformation has been found useful. Its use cannot be rationalized in the same way as the logarithmic transformation but it seems usually to apply to total building costs, as distinct from costs per unit area.

As with any transformation, an indication for its use can be obtained by applying it first to some of the quantiles of the distribution and noticing whether they become symmetrical about the transformed median.

As with other transformations, if the mean and standard deviation are required for quotation, rather than for comparison with others, they should be calculated from the untransformed data.

4.6 The normal distribution

Populations of price data, if symmetrically distributed, or after transformation to symmetry, usually conform to the normal distribution. This means that, if their arithmetic mean is subtracted from each item so that they become deviations from the mean, and if these are divided by their standard deviation, the resulting distribution always has the same characteristic shape. The adjustments are called 'standardization'. They make the mean zero and the standard deviation unity. This distribution is tabulated in several ways and at least one version is printed in most statistics books and all books of statistical tables.

In a smooth frequency distribution or frequency curve, the height at any value band on the horizontal axis is proportional to the number of cases within that value band. This can be thought of as the height of a narrow strip. Then the number of cases between any two specified values is proportional to the total of the strips making up the area under the curve between these two values. Thus frequency is proportional to area in a frequency distribution.

Figure 4.4 shows the shape of the standardized normal distribution and notes some useful proportions. Because it is symmetrical the proportion of cases greater than the arithmetic mean is 1/2. The proportion more than one standard deviation above the mean is about 1/6. The proportion more than two standard deviations above the mean is about 1/40. The distribution is symmetrical, so the same proportions apply below the mean.

It is worth committing to memory the proportion of cases within two standard deviations above or below the arithmetic mean. It is 95%. Also, 2/3 are within one standard deviation. These proportions are noted in Fig. 4.4. An alternative to the last is that 1/2 the cases are within 2/3 of a standard deviation of the mean.

One reason for memorizing the proportions in Fig. 4.4 is that it is easy to misinterpret tables when they are presented differently from what was

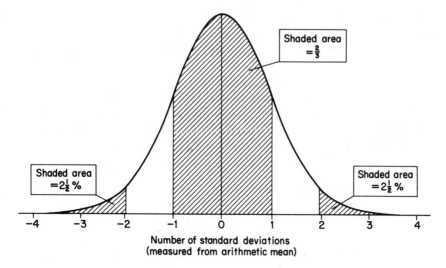

Fig. 4.4 Approximate proportions of area under the normal curve between or outside specified ordinates.

expected. Rough checks of the above proportions in the table confirm that it is being correctly read.

4.7 Tables of the normal distribution

There are four basic ways in which the proportions of area under the curve are commonly related to the number of standard deviations above the arithmetic mean.

The first gives the proportion of area under the curve to the right of, therefore more than, tabulated points. These proportions are sometimes expressed as percentages and sometimes as decimal fractions. They are often called probabilities because of the purpose for which the proportion is usually required. This is dealt with in Section 5.3 which deals with testing significance. The proportion to the right of one standard deviation greater than the mean is 0.159 so the $\frac{1}{2} - \frac{1}{3}$ which could be deduced from Fig. 4.4 is not exact.

The second method of tabulation inverts the above by giving the number of standard deviations from the mean corresponding to specified areas.

The third method does not distinguish between positive and negative deviations. It tabulates the proportion of the area under the normal curve outside specified numbers of standard deviations from the mean. In such tables a proportion of about 1/20 (0.046 in fact) is shown as being more than two standard deviations from the mean. It is between this type of table and the first type that the commonest confusion is made. The first type is often called 'single-tailed' and this 'double-tailed'. Mistakes can easily be avoided if the rough proportions given earlier are remembered.

The fourth method is an inversion of the third, in the same way as the second is an inversion of the first.

Often only one sort of table is available when another sort is required. Using the second or fourth methods backwards to give the first or third is obviously easy but to use a single-tailed table when a double-tailed type is required is less easy and needs care. It is helpful to draw rough diagrams shaded as in Fig. 4.5 and labelled as required.

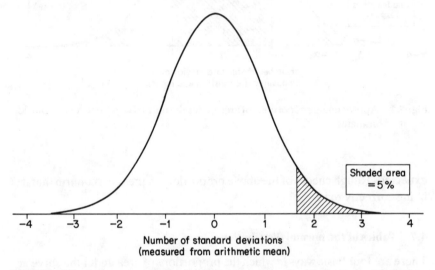

Fig. 4.5 5% of the area under the normal curve is more than 1.64 standard deviations above the arithmetic mean.

Table A at the end of the book is of the single-tailed type giving the proportion of the area under the curve to the right of various numbers of standard deviations above the mean. Only half the distribution is tabulated. If required, the other half can be deduced by symmetry.

Some calculators include the area under the normal curve as a function, thus providing an alternative to the use of tables. They are quicker to use than tables but the time which can be saved in this way is very little and the facility is not required often enough to justify a large increase in the cost of the calculator. On the other hand, mistakes are less likely and tables are not always to hand.

Tables of areas under the normal curve must not be confused with tables of ordinates of the normal curve. These give the height of the curve at any point and are only required if the curve is to be plotted.

4.8 Using the normal curve

The normal curve is usually described as 'bell-shaped'. This can be misleading because, although its height is very low beyond about $2\frac{1}{2}$ standard deviations from the mean, there is no point at which it can truly be said to be zero. A table with five places of decimal shows that beyond even four standard deviations above the mean there is still 0.003% of the total area under the curve.

If the arithmetic mean and standard deviation of a population of prices is known, the reasonableness of looking upon a particular price as belonging to the population can be estimated. If it is four standard deviations above the mean, the 0.003% probability (single-tailed) of finding such a high or higher price in the population can, for practical decision making, be regarded as impossible. Even the 1% probability of finding cases more than 2.6 standard deviations greater or less than the mean (double-tailed) is unlikely enough to be ruled out for nearly all purposes. In fact, where a decision must be made one way or the other using the only evidence available, much higher probabilities can form the basis for action. This is more fully discussed in the next chapter.

Deductions of this sort are so useful that the reasoning behind them, described above, needs to be thoroughly understood and practised. It forms the basis of the significance tests to be explained later (Chapter 5) and leads on to many more statistical procedures, provided that it can be assumed that the shape of the distribution is approximately the same as normal.

Many populations, though not exactly normally distributed, are close enough for them to be treated as though they were. Experience with analysis of price data soon shows the extent of the departures from normality which can be accepted before the assumption of normality becomes misleading. For many purposes it is surprisingly great.

Even moderate skewness can be accepted if the double-tailed approach is applicable. The proportion of area in one tail can be very different from the tabulated figure but the two tails together often agree well enough for many purposes. With small samples, when nothing is known about the population, conclusions have to be very tentative so the error introduced by the unjustified assumption of normality may be unimportant. Sweeping assumptions made with the eyes open are less dangerous than some of those which are implied by some procedures but not understood.

The advantages of assuming normality are great because by calculating only two parameters, the arithmetic mean and the standard deviation, the whole population distribution is 'known'. Because of this some researchers in the past have been guilty of mistaken conclusions based upon unjustified assumptions of normality. This is less likely in the analysis of building price data than in most fields.

Data which conform to the normal distribution have a tendency towards a common value, the mean, and the individual items can be regarded as deviations from it. This is most clearly the case when there is some sort of target and each item deviates from it through error. In fact, an early name for the normal distribution was 'the error distribution'.

A distribution of numbers which vary because of errors can only be truly normal if it is symmetrical, and this requires that errors above the mean do not tend to be greater or smaller than errors below the mean. Whether the mean error is positive or negative is not important.

It is interesting to produce normal distributions by experiment. Demonstrations commonly used by the author with more than about 40 students are to ask them to estimate the height of Nelson's Column from memory; the width of the room; and to measure their pulse rate. The first of these produces a skewed distribution with one or two wildly high figures. However, the median is usually surprisingly close to the true figure (170 feet or 52 metres). The other two distributions are usually normal. The median and arithmetic mean of the room widths are always very close to the true figure. When one of the sleepier students has misunderstood the task and recorded a wildly wrong figure, the median shows its superiority over the arithmetic mean.

A direct demonstration of a normal distribution is provided by physically aiming at a target. Throwing large numbers of balls of paper at a target low down on a wall and sweeping them towards the wall produces a normal distribution directly.

It is not surprising that estimates and estimating errors, when expressed as proportions of the target, are normally distributed. The distribution of errors is usually centred a little above zero and some skewness can sometimes be discerned. These effects result from a wish on the part of the estimator to avoid large errors in one direction even at the risk of increasing the tendency to error in the other direction. However, the skewness produced is only slight and can safely be ignored. The effect on the mean error is also not great but need not be ignored because it is easy to deal with in any analysis.

4.9 Differences in patterns (the χ^2 test)

When a frequency distribution is being appraised, the question arises of whether differences from what had been expected indicate a genuine difference or could easily have arisen by chance. There is a numerical test which assists in this judgement by evaluating the probability that differences of the size of those being observed, or greater differences, would have been produced if the differences had occurred randomly.

This test is the χ^2 test (the Greek letter *chi* usually pronounced 'ky' as in 'sky') and is concerned with observed and expected frequencies. The expected frequencies could have come from a variety of sources. The comparison of a frequency distribution with what would be expected if the data were normally distributed is used in Tables 4.1 and 4.2 as an example of the use of the χ^2 test.

The most difficult stage is the calculation of expected frequencies in Table 4.1. The distribution must first be standardized. This requires the cell boundaries to be expressed as the number of standard deviations they are from the arithmetic mean. These standardized cell boundaries are in the third column of Table 4.1. The population standard deviation and arithmetic mean are better estimated from the basic data than from the frequency distribution.

Table 4.1

Lower cell boundaries	Frequency	Lower cell boundaries as number of s.d.s from a.m.	Proportion of area under normal curve	Expected frequency (proportion × 50)
$-\infty-$	0	$-\infty\sigma-$	0.04	2.0
20–	4	$-1.8\sigma-$	0.16 – 0.04 = 0.12	6.0
30–	18	$-1.0\sigma-$	0.42 – 0.16 = 0.26	13.0
40–	17	$-0.2\sigma-$	0.69 – 0.42 = 0.27	13.5
50–	3	$+0.5\sigma-$	0.90 – 0.69 = 0.21	10.5
60–	6	$+1.3\sigma-$	0.98 – 0.90 = 0.08	4.0
70–	2	$+2.1\sigma-$	1.00 – 0.98 = 0.02	1.0
80–	0	$+2.8\sigma-$	0	0
	50			50

Arithmetic mean = 43.0

The sign ∞ at the top of the first and third columns means infinity, σ represents the standard deviation (s.d.) and a.m. the arithmetic mean. The proportions of area in the fourth column are the differences between the entries in Table A corresponding to the boundaries in the third column.

The first point to notice is that the expected frequencies must add up to the same as the observed. If they had not added up exactly, scaling all up or down would have been required. Next, although in this example whole numbers would give sufficient accuracy, decimal fractions are required in the expected frequencies unless all frequencies are greater than about 10.

Before the expected and observed frequencies can be compared, a condition of the validity of the χ^2 test must be satisfied. This is that the expected frequencies must not be too small. An acceptable minimum is 5. To achieve this, cells must be grouped as in Table 4.2. There is no need to keep cell widths the same.

The value of χ^2 is the sum of the contributions made by the individual cells. This contribution is calculated by squaring the difference between the observed and expected frequencies and dividing by the expected frequency.

Table D at the end of the book gives the probability that such a value of χ^2 or a higher one would be produced if the sample had been randomly drawn from a normally distributed population. To use the table, the number of degrees of freedom must be decided. This is the number of pairs of frequencies being compared (in this case 5) minus the number of parameters which had to be calculated from the data to determine the expected frequencies. In this example the mean and the standard deviation were needed. Thus the number of degrees of freedom is 3 and the probability that this value of $\chi^2 = 12.0$ could have occurred by chance if the sample came from a normal population is given by Table D as slightly less than 1% (probability = 0.01).

Table 4.2

Lower cell boundary	Observed frequency (O)	Expected frequency (E)	$\dfrac{(O-E)^2}{E}$
0–	4	8.0	2.0
30–	18	13.0	1.9
40–	17	13.5	0.9
50–	3	10.5	5.4
60–	8	5.0	1.8
			$\chi^2 = 12.0$

Most people look upon a 1% probability as highly unlikely and would reject a hypothesis that the sample came from a normal population.

A χ^2 value as low as 7.8 would have been required for the probability to have been increased even to 5%. It is clear from the contributions to χ^2 in Table 4.2 that the main culprit was the 50 to 60 group.

It should be noted that the χ^2 test is not concerned with the sequence of the frequencies, so a poor fit caused by skewness, as this one probably is, is not distinguished from erratic departures with no special pattern.

The next example of the application of the χ^2 test is the comparison of two samples when there is no external basis for the calculation of expected frequencies. The comparison is required to help assess whether they might both have come from the same population.

In this case the hypothesis to be tested is that both samples are from the same population. If this is so then the two samples can be combined to give a better picture of the population. The expected frequencies for each sample can be calculated from the combined frequencies. The first step is to obtain the combined proportions as shown in Table 4.3.

Table 4.3

Lower cell boundary	Sample A frequencies	Sample B frequencies	Combined sample frequencies	Combined sample proportions
30–	1	4	5	0.10
35–	6	4	10	0.20
40–	11	2	13	0.26
45–	5	5	10	0.20
50–	8	1	9	0.18
55–	2	1	3	0.06
60–	0	0	0	0
Total	33	17	50	1.00

The last column provides expected frequencies for each of the samples A and B if it is assumed that the proportions obtained from the combined sample apply to each of the two samples individually. In practice the figures in the last column in Table 4.3 need not be written down. They can be held one at a time in the calculator while the expected frequencies in Table 4.4 are obtained directly from the combined sample frequencies by simple proportion.

Table 4.4

Lower cell boundary	Expected frequencies for sample A	Expected frequencies for sample B
30–	3.3	1.7
35–	6.6	3.4
40–	8.6	4.4
45–	6.6	3.4
50–	5.9	3.1
55–	2.0	1.0
60–	0	0
	33.0	17.0

Strict application of the rule that the minimum expected frequency should be no less than 5 would require the cells to be collapsed to 2. A comparison based on only 2 cells may hardly be worth making on its own so in this example the rule is stretched a little. The results must therefore be used with caution. The contribution to χ^2 of the deficient cell must be specially noted in case it is large and ought to cast doubt on the answer. χ^2 is calculated in Table 4.5.

Table 4.5

Lower cell boundary	Frequencies for sample A Exp.	Obs.	Contributions to χ^2 in sample A	Frequencies for sample B Exp.	Obs.	Contributions to χ^2 in sample B
30–	9.9	7	0.85	5.1	8	1.64
40–	8.6	11	0.23	4.4	2	1.31
45–	14.5	15	0.02	7.5	7	0.03
	33.0	33	1.10	17.0	17	2.98

χ^2 from A and B = 4.1 (of which 1.3 comes from the deficient cell)

The table of χ^2, Table D, will be used to find the probability that this or a greater value of χ^2 could have occurred by chance if both frequency distributions represented samples from the same population. To use the table, the number of degrees of freedom must first be worked out.

The number of degrees of freedom equals the number of cells in the table of observed frequencies to which frequencies could be allocated freely without affecting the marginal totals. This is difficult to explain except by example. In Table 4.6 the marginal totals of the observed frequencies have been added.

Table 4.6

Lower cell boundary	Observed frequencies		
	Sample A	Sample B	Total
30–	7	8	15
40–	11	2	13
45–	15	7	22
	33	17	50

Without changing any of the totals it would be possible to change the 7 and 11 in Sample A if compensating changes were made in the other four figures. This applies to any two figures in the table, so the number of degrees of freedom is 2.

When expected frequencies are obtained by combining samples, the number of degrees of freedom is 1 less than the number of columns multiplied by 1 less than the number of rows; in this case $(2-1) \times (3-1)$.

Table D shows the probability that a value of χ^2 of 4.1 or more could have been produced by chance. For 2 degrees of freedom it is about 0.15 (15%). This is too high to be taken as good evidence that the hypothesis that the two samples came from the same population is untrue.

If the number of degrees of freedom is only 1, a correction to the calculated value of χ^2 is required to allow for the fact that the observed frequencies can only be whole numbers whereas the expected frequencies are fractional. Rather than rounding the expected frequencies to whole numbers a continuity correction should be applied by reducing the difference between the observed and expected frequencies by 0.5.

The following example of a continuity correction using the data in Table 4.7 also shows how the χ^2 test is applicable to any classified data, not just those arranged in frequency distributions.

Table 4.7 Repair data from storage buildings for two types of roof (expected frequencies in brackets)

Type of roof	Repaired in first 10 years	Not repaired in first 10 years	Totals
Pitched	16 (18.4)	19 (16.6)	35
Flat	15 (12.6)	9 (11.4)	24
	31	28	59

The expected frequencies for pitched roofs were calculated by taking the overall proportion of pitched roofs (35/59 = 0.5932) and applying it to each of the foot-totals (31 and 28). The others were obtained by subtraction from the foot-totals.

Note that, in a two-by-two table, the four differences between the observed and expected frequencies are all the same. In this case they are 2.4. The continuity correction reduces this to 1.9 so the sum of the contributions to χ^2 by the four cells is

$$1.9^2/18.4 + 1.9^2/16.6 + 1.9^2/12.6 + 1.9^2/11.4$$

This is easier to calculate as

$$1.9^2(1/18.4 + 1/16.6 + 1/12.6 + 1/11.4) = 1.9^2 \times 0.2817 = 1.0$$

Table D shows that for one degree of freedom a value of χ^2 as great as this could occur by chance with a probability of about 30%. This is much too high for the data to be taken as evidence of a genuine difference between the two types of roof in this respect. Of course, this is not to say that there is no difference but simply that these data do not indicate one.

5 Confidence in estimates

5.1 Confidence Limits

When all efforts have been made to ensure that a sample is as representative as possible, estimates of the parameters of the population which the sample is intended to represent are calculated from the sample.

For example, if a cost per unit floor area for a type of building is being estimated, cases conforming to the desired description are drawn from a data bank and their average cost per unit area used as an estimate of the likely cost per unit area of such buildings. It may be that the design for which an estimate is required is expected to cost more than the average price so, as well as the population average, a measure of the spread is needed. Then the estimate can be pitched at an appropriate place in the distribution. The arithmetic mean and standard deviation are two of the population parameters commonly needed.

Because they are calculated from a sample, the estimates of the required population parameters cannot be regarded as firm. Other samples could have arisen by chance which would have led to different estimates of the population parameters.

For this and other applications it is useful to have an idea of how good the estimates of the population parameters are likely to be. Although this cannot be assessed with certainty, limits can be set which are likely to be exceeded with only a small, stated probability. If this probability is, say, 1% it could be said, in non-statistical terms, that we are 99% confident that the true answer lies within the limits we have set. Such limits are called *confidence limits*.

It is easy to guess that the larger the sample, the narrower the confidence limits. Also the smaller the spread in the population, the narrower the confidence limits.

The following sections cover the setting of confidence limits for the parameters which are of most use in the analysis of price data and estimating.

5.1.1 Confidence limits for the arithmetic mean

A sample of one
If the sample of price data consists of only one item, an attempt should be made to imagine the distribution of the parent population to get an idea of

what the single item might have been if chance had selected a different item from the population. It is essential to the statistical appreciation of data to think in this way rather than looking upon the item of data as unique.

The item may be the only guide available to the 'true price' which, it may have been decided, is the arithmetic mean of the parent population; but the item could have come from anywhere in the distribution of the parent population. Although it is obviously more likely to have come from somewhere in the middle than from anywhere else in the distribution, it is still possible that it came from one of the tails. If it had happened to come from one of the tails which each account for $2\frac{1}{2}\%$ of the population, the population mean would have been two standard deviations or more above or below the single item price. Imagine the population distribution sliding from left to right in Fig. 5.1.

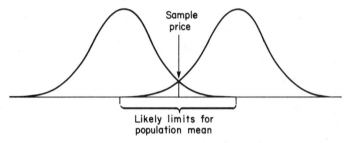

Fig. 5.1 Two extreme positions for the population frequency distribution.

There is a 5% probability that the sample item came from one or other of these tails and a 95% probability that it did not. These limits (approximately ± 2 standard deviations from the sample item) are called 95% confidence limits for the population arithmetic mean, because 95% can be quoted as a measure of the confidence that we have that the population mean lies within the limits.

Confidence limits based on a sample of one can only be calculated if we already know the population standard deviation or can estimate it from analogous data. Luckily, different populations in the same field may have different means but similar standard deviations or, more likely, similar coefficients of variation. With familiarity with the type of data being analysed, it is often possible to make quite a good assessment of the population standard deviation.

Routine statistical analysis of data provides this feeling for variability which aids judgement of the likely standard deviation. This deliberate development of statistical judgement can be made a quicker and more organized process than the unstructured development of professional or 'seat of the pants' judgement of prices but it is of the same nature and, properly planned, can provide a rapid way of equalling the performance of the most experienced in the field. Also, statistically based judgement is easier to transfer to others, provided that all have some knowledge of statistical methods to provide a common language.

No matter how much analysis has been done and judgement acquired, an estimated population standard deviation is still only an estimate. Some allowance for this must be made when setting the confidence limits. How much this allowance should be depends on the soundness of the basis of the estimate of the population standard deviation.

A feeling for how large this factor should be can be gained from the next section where an analogous factor (t) is applied to reflect the reliance which can be placed on estimates of population parameters made from samples of various sizes.

Samples larger than one
It is to be hoped that reliance seldom has to be placed on samples smaller than 5, although in common practice it is widely believed that carefully choosing a single item of data provides a better estimate of the 'true price' (i.e. the population mean) than averaging several. Perhaps those who have read so far will think otherwise when they consider how much variability is unassigned and therefore effectively random.

For the following theory, the concept of repeatedly sampling at random from the same population is required. The standard deviation of the means of all possible samples of a given size is less than the standard deviation of the population, whereas for a sample of one it was equal to it. The standard deviation of the sample means is inversely proportional to the square root of the sample size. It is given by

$$\frac{\sigma}{\sqrt{n}}$$

where σ is the population standard deviation and n the sample size. This standard deviation of the arithmetic mean is called its *standard error*. Every parameter has its standard error.

The distribution of the arithmetic means of all possible samples from a population is approximately normal for sample sizes greater than 4 and closely follows the normal distribution for large samples. Amazingly, this is true even if the population distribution is not normal.

The proof of this is known as the central limit theorem, but the reader can demonstrate it by taking samples of, say, 10 random numbers and plotting the distribution of their means (or totals). This is easily done with a programmable calculator which has a random number generator. The shape of the distribution of random numbers is rectangular.

The population standard deviation is usually unknown but, if the sample size is larger than 4, it should be capable of providing a reasonably good estimate of it. This is obtained by applying Bessel's correction to the standard deviation of the sample. For a sample of size n,

$$\text{Estimated population s.d.} = \text{sample s.d.} \times \sqrt{\left(\frac{n}{n-1}\right)}$$

This is often given directly by the σ button on a calculator which has a suffix $(n-1)$.

Although this estimate is unbiased, it is nevertheless only an estimate, so an extra allowance must be made when using it to set confidence limits. They must be widened according to the size of the sample upon which the estimate of the population standard deviation is based. The smaller the sample, the wider the limits.

The confidence limits also vary according to the level of confidence required. Both the adjustment for sample size and that for confidence level are included in one factor (t) which is tabulated for various sample sizes and confidence levels. In fact the tables are not based on sample size directly but on degrees of freedom, which in this application are one less than the sample size.

A version of the table is included at the end of this book (Table C). It is used by looking down the column appropriate to the level of confidence being used (double-tailed). The required value of t is against the degrees of freedom.

The confidence limits for a parameter are above and below the parameter calculated from the sample. The amount by which they are above and below is $t \times$ (estimated standard error).

For example, if the standard deviation of a sample is 10.5 and there are 6 items in it the estimated standard deviation of the population is

$$10.5 \sqrt{(6/5)}$$

The estimated standard error of the mean is

$$\frac{10.5 \sqrt{(6/5)}}{\sqrt{6}} = \frac{10.5}{\sqrt{5}} = 4.70$$

When this is thoroughly understood it is safe to omit the step involving Bessel's correction, because $\sqrt{n}$ cancels out ($\sqrt{6}$ in the above example), and go straight to calculating the estimated standard error of the mean by dividing the standard deviation of the sample by $\sqrt{(n-1)}$.

The value of t for 5 degrees of freedom and two-tailed 95% probability is 2.571, so the double-sided confidence limits are $2.571 \times 4.70 = 12.07$ above and below the arithmetic mean of the sample. If the mean is 40, the 95% confidence limits are 28 and 52.

These limits indicate that we can be fairly sure that the 'true mean' (the population mean) lies between 28 and 52. For many purposes these limits would be too wide so a larger sample would be called for.

In some circumstances the size of the population is small and the sample is a considerable proportion of it. If the sample size were increased to approach the population size the confidence limits should narrow until they coincide. To produce this narrowing, a *finite population correction* is applied instead of Bessel's correction. The standard error of the mean becomes

$$\left(\frac{\sigma}{\sqrt{n}}\right)\sqrt{\left(\frac{N-n}{N-1}\right)}$$

where n is the sample size and N is the population size. Thus when n and N are equal the standard error becomes zero.

5.1.2 Blending estimates of the population standard deviation

If an estimate of the population standard deviation is already available from other work, it should be blended with the estimate from the current sample. With experience this can often be done by using judgement, but judgement can only be developed by experience of the results of calculation.

Blending is especially important when the current data are a sample of fewer than 5 items. The estimate of a population parameter which they provide is of little value on its own but can be a valuable addition to a previous estimate.

If the previous estimate of the population standard deviation was also from a sample and the old and new data are of roughly equal credibility, the blending can be done as follows. If the standard deviations of the two samples are s_1 and s_2 respectively and their sample sizes are n_1 and n_2, the estimated population standard deviation is

$$\sqrt{\left(\frac{n_1 s_1^2 + n_2 s_2^2}{n_1 + n_2 - 2}\right)}$$

The following example shows how the combined estimate compares with its two components.

If two samples of 10 and 4 items had standard deviations of 9.0 and 12.0 respectively, their separate estimates of the population standard deviation would, using Bessel's correction, be

$$9.0 \sqrt{(10/9)} = 9.5$$

and

$$12.0 \sqrt{(4/3)} = 13.9$$

It might be guessed that the combined estimate would be close to the mean of the individual estimates weighted by sample size. This is

$$\frac{10 \times 9.5 + 4 \times 13.9}{14} = 10.8$$

A weighted mean is easy to guess.

In fact the combined estimate is

$$\sqrt{\left(\frac{10\times 9.0^2 + 4\times 12.0^2}{10+4-2}\right)} = 10.7$$

so a guess at the weighted mean would have been close enough.

The ability to guess well should be developed for use when full calculation is not worth while and a roughly approximate answer will do.

If more confidence can be placed in one sample than the other, the weights in a guess, or the sample sizes in the calculation, can be altered to reflect this. For instance, it may be felt that one item in the new sample is worth two in the old. In that case, when combining them, the pseudo sample sizes can be made $n_1 = 5$ and $n_2 = 4$. The combined estimate of the population standard deviation then becomes

$$\sqrt{\left(\frac{5\times 9.0^2 + 4\times 12.0^2}{5+4-2}\right)} = 11.8$$

Again, a guess at the weighted mean would probably have been adequate. The weighted mean is

$$\frac{5\times 9.5 + 4\times 13.9}{9} = 11.5$$

The same principles can be applied when there are more than two sources of information to be blended. For three samples, their combined estimate of the population standard deviation is

$$\sqrt{\left(\frac{n_1 s_1^2 + n_2 s_2^2 + n_3 s_3^2}{n_1 + n_2 + n_3 - 3}\right)}$$

5.1.3 Confidence limits for the median

The standard error of the median of an approximately normal distribution is 1.25 times the standard error of the arithmetic mean.

If the median is being used, it is possible that the standard deviation of the sample will not have been calculated but that the interquartile range (IQR) is known. It is, of course, the interval between the quartiles. In that case the population standard deviation can be estimated as $\frac{3}{4}$ of the sample interquartile range multiplied by Bessel's correction,

$$\text{i.e. } 0.75 \times \text{IQR} \sqrt{\left(\frac{n}{n-1}\right)}$$

If the mean deviation (m.d.) of the sample has been calculated, the population standard deviation can be estimated by multiplying it by $1\frac{1}{4}$ and applying Bessel's correction,

$$\text{i.e. } 1.25 \text{ m d} \sqrt{\left(\frac{n}{n-1}\right)}$$

These relationships are strictly true only when the population is normally distributed, but they are not greatly affected by moderate departures from normality. The setting of confidence limits for building price data cannot be so precise that such departures are important, but if it is thought desirable to make adjustments a little experimental analysis of skew distributions will give an idea of how sensitive the relationships are to skewness in the distribution.

When the population standard deviation (σ) has been estimated, the standard error of the median for a sample size n can be calculated. It is

$$\frac{1.25\sigma}{\sqrt{n}}$$

This is 25% greater than the standard error of the arithmetic mean and, with well-behaved data, this is a fair measure of the relative merits of the mean and median. However, the median is less affected by erratic data which contain the sort of errors (mistakes) not measured by the standard error so, for the analysis of building price data, this cancels out the theoretical disadvantage.

When the standard error of the median has been calculated, the confidence limits for the population median can be set in the same way as for the mean. They are

$$\text{Sample median} \pm \frac{1.25t\sigma}{\sqrt{n}}$$

where the value of t is obtained from Table C.

5.1.4 Confidence limits for other parameters

It is possible to calculate confidence limits for any statistical parameter. The method is the same as for the mean and median once the standard error is known. The amount by which the limits are above and below the value of the parameter calculated from the sample is t multiplied by the standard error; t is read from Table C for the chosen level of confidence and with degrees of freedom one less than the sample size.

The standard error is calculated from the sample size (n) and the population standard deviation (σ). When the population standard deviation is estimated from the sample, Bessel's correction must first be applied.

The standard errors for the parameters most useful in the analysis of building price data are shown in Table 5.1.

Table 5.1

Parameter	Standard error
Arithmetic mean	$\sigma/\sqrt{n}$
Median	$1.25\sigma/\sqrt{n}$
Standard deviation	$\sigma/\sqrt{(2n)} = 0.71\sigma/\sqrt{n}$
Mean deviation	$0.60\sigma/\sqrt{n}$
Semi-interquartile range	$0.79\sigma/\sqrt{n}$
Coefficient of variation	$\dfrac{V}{\sqrt{(2n)}}\sqrt{\left(1+\dfrac{2V^2}{10\,000}\right)} =$ approx. $V/\sqrt{(2n)}$ if $V < 50\%$
Quartiles	$1.36\sigma/\sqrt{n}$
Deciles (first and ninth)	$1.71\sigma/\sqrt{n}$

The coefficients of $\sigma/\sqrt{n}$ for all but the lowest few percentiles are given by $1.12 + 6/P$, where P is the percentile (or 100 minus the percentile if it is greater than 50%). For instance, P for the quartiles is 25, so the standard error of a quartile is $(1.12 + 6/25)\sigma/\sqrt{n} = 1.36\sigma/\sqrt{n}$.

5.1.5 Quoting confidence limits

In most statistical work, by far the commonest confidence limits used are 95%. This is because a 5% chance of being wrong is plenty low enough for most purposes. Where greater certainty is required, because important decisions depend upon the answer, it is usual to quote 99% confidence limits.

In building, the importance of the decision hanging on the interpretation of numerical data varies tremendously. If there is doubt about which probability level is appropriate, it is wise to go for wider rather than narrower limits. The gain in credibility represented by the choice of 99% confidence limits rather than 95% is often well worth the widening of the limits. For instance, for sample size 6 the limits for the arithmetic mean have only to be widened from $\pm 1.0\sigma$ to $\pm 1.6\sigma$ to widen the double-tailed confidence limits from 95% to 99%. However, 99.9% confidence limits are at $\pm 2.8\sigma$ for sample size 6. Such wide limits are seldom of much practical use so this level of confidence can only be obtained from large samples.

5.1.6 Asymmetrical confidence limits

Where the population distribution is skewed and the sample is much less than about 30, the confidence limits should, strictly speaking, be asymmetrical. If the skewness is to the right, the upper limit should be further above the average than the lower one is below.

However, the asymmetry in the confidence limits should be much less than the skewness in the population distribution even for samples sizes as small as 5, and the limits become more symmetrical as sample size increases. It may be sufficient to adjust the confidence limits subjectively but, if so, the adjustments should be kept small. If they are made in this way, both limits should be raised but the upper limit more than the lower.

If the transformation which would normalize the population distribution is known and it is desired to use it before calculating confidence limits, it is safer to draw conclusions from the confidence limits without retransforming to the original units. Retransformation can be done for presentation purposes, preferably by simply retransforming the scale of a graph of the transformed data.

5.1.7 Single-sided confidence limits

So far it has been assumed that confidence limits are required giving a stated probability that the chosen parameter will lie between them. This is not always what is wanted. Sometimes there is no interest in one of the limits so it is enough to give just the other one. This is usually the upper. The stated probability then refers to the probability of the population parameter being below the single confidence limit.

Such a limit is called a single-sided confidence limit. It is calculated in a similar way to double-sided confidence limits; the only difference in the case of a sample of one is that the probability is considered of the item having come from a tail of the distribution accounting for 5% of the population. This is only 1.6 standard deviations from the mean instead of about 2. For samples larger than one, when the table of t is entered, the single-tailed probabilities are used.

Thus, in the example in Section 5.1.1, if the object of sampling was only to decide on a maximum figure for the population arithmetic mean the single-sided confidence limit would be calculated as follows. The value of t for 5 degrees of freedom and a single-tailed probability of 95% is 2.015 so there is a 95% probability that the population mean does not exceed $20 + 2.015 \times 4.70 = 29$.

5.2 Deciding sample sizes

The procedure for setting confidence limits can be used in reverse to answer one of the questions most frequently asked of statisticians. 'How big a sample is needed to estimate the true value?'

This question is too vague to be answered as it stands but can be turned into one which can be answered such as, 'What sample size will give, say, 95%

confidence that the population median lies within, say, 30% of the median of the sample?'

Call the median of the sample M and the minimum sample size required n. The first step is to turn the confidence limits specified into a required standard error as follows.

The 95% confidence limits specified are $M \pm 0.3M$. The standard error of the median is given in Table 5.1 as $1.25\sigma/\sqrt{n}$, so the confidence limits are

$$M \pm t \times \frac{1.25\sigma}{\sqrt{n}}$$

Thus $$t \times \frac{1.25\sigma}{\sqrt{n}} = 0.3M$$

so $$\frac{t}{\sqrt{n}} = \frac{0.3}{1.25} \frac{M}{\sigma}$$

Next, the population standard deviation must be estimated as a proportion of the median. If other analyses of populations of this type have indicated a coefficient of variation of about 25%, this can be taken to be an estimate of σ/M. Therefore M/σ is estimated as 4, so

$$\frac{t}{\sqrt{n}} = \frac{0.3}{1.25} \times 4 = 0.96$$

Lastly, the 95% double-tailed column of the table of t (Table C) is searched for a pair of values of $(n-1)$ degrees of freedom and t (double-tailed) which gives $t/\sqrt{n}$ approximately equal to 0.96. This is nearest for $n = 7$ where $t = 2.447$ so $t/\sqrt{n} = 0.92$.

The indicated sample size is thus 7, but it would be unwise to recommend it before calculating the effect of a wrong assumption of the population standard deviation. If it should have been $\frac{1}{3}$ of the median rather than $\frac{1}{4}$,

$$\frac{t}{\sqrt{n}} = \frac{0.3}{1.25} \times 3 = 0.72$$

This requires a sample size of 10 so the sample size is quite sensitive to the accuracy of the estimate of the population standard deviation.

If a sample size of 7 were to be used when it should have been 10, the 95% confidence limits produced would be in the region of

$$M \pm t \times \frac{1.25\sigma}{\sqrt{n}}$$

$$= M \pm 2.447 \times \frac{1.25}{\sqrt{7}} \times \frac{M}{3}$$

$$= M \pm 0.39M$$

instead of the $M \pm 0.30M$ which had been aimed for.

The sample size actually used will depend on three things: the importance of keeping the confidence limits to the specified level, the cost of sampling and analysis, and the difficulty in going back to take another sample if the first proved to have been too small. If the specified confidence limits were intended to be the widest admissible, it would be necessary to use a sample size of at least 10, even if the population standard deviation were accurately known. This is to allow for the estimate of the population standard deviation obtained from the sample being as high as $0.33M$. This could easily be the case because the standard error of the standard deviation is $\sigma/\sqrt{(2n)}$. If the true population standard deviation were $0.25M$, the standard error in samples of size 10 would be

$$\frac{0.25M}{\sqrt{20}} = 0.056M$$

Thus $0.33M$ is only 1.4 standard errors above $0.25M$. It might be thought wiser to calculate a sample size on the assumption that it could be as high as 2.2 standard errors (t with between 10 and 15 degrees of freedom) above the median. This would be $(0.25 + 0.12)M = 0.37M$ and lead to a recommended sample size of 12.

It is not surprising that setting sample sizes is not a precise process, bearing in mind the element of chance in selecting a sample, but calculating likely confidence limits brings some rationality to what would otherwise be pure guesswork.

5.3 Testing significance

The distribution of the factor t was tabulated by W. S. Gosset who wrote under the name 'Student'. It provides one of the most commonly used significance tests. One significance test has already been described – the χ^2 test. This one is called the t test and is most commonly concerned with either the difference between the arithmetic mean of a sample and that of a population, or the difference between two sample means.

The principle of a significance test is that of testing a hypothesis. For instance, the hypothesis would be that a sample has been randomly drawn from a specified population. In another type of problem the hypothesis could be that two samples were randomly drawn from one population. The significance test in each case consists of calculating the probability that such samples, or more different ones still, could have been randomly drawn and therefore that the differences between means is merely the result of chance. If this probability is too small the hypothesis is rejected. These hypotheses are stating that apparent differences are due to chance. They are called *null hypotheses*.

The most sensible way for those dealing with building price data to use significance tests is not in this accept-or-reject way, but to take the value of the

Confidence in estimates

calculated probability into account when forming a judgement, making a decision or planning action. This is easy enough if the basis of the test is understood but is too difficult to convey to those without this understanding. This makes it important for those making decisions based on the statistical analysis of data to understand the statistical techniques used.

5.3.1 Student's t test

To calculate the probability that the amount by which the arithmetic mean of a sample differs from the population mean could be accounted for by chance requires the estimation of the population standard deviation. The standard deviation of the sample provides such an estimate by the use of Bessel's correction, as in the following example.

An estimator claims that his average estimating error (regardless of direction) is 10%. Are the results from his last 10 estimates compatible with this claim? They are 5%, 30%, 14%, 41%, 3%, 8%, 31%, 28%, 19%, 25%.

The null hypothesis is that these 10 results are random drawings from a population of all his possible estimating errors, whose arithmetic mean is 10% and standard deviation unknown.

The arithmetic mean of his errors is 20.4% and their standard deviation is 12.01%. The estimate which they provide of the population standard deviation is

$$12.01 \sqrt{\left(\frac{10}{9}\right)} = 12.7\%$$

The standard error of the mean is

$$12.7 \div \sqrt{10} = 4.0$$

Therefore the number of standard errors which the sample mean is above the mean of the hypothetical population is

$$\frac{20.4 - 10}{4.0} = 2.6 \text{ standard errors}$$

The table of t (Table C) is consulted to find the probability that a sample mean at least 2.6 standard errors from the population mean could be randomly drawn from the hypothetical population. The number of degrees of freedom is one less than the number of items in the sample, i.e. 9, so from Table C the probability (double-tailed) is about 0.03. This is so unlikely that the estimator's claim can reasonably be disbelieved although no important decision should be taken based on the disbelief alone. A threshold is not necessary, but if one is desired, the lowest probability conventionally acceptable is 0.05 and this corresponds to 2.26 standard errors from the population mean. Thus his average performance would have needed to be within $2.26 \times 4.0 = 9.0\%$ of the claimed 10% (i.e. between 1.0% and 19.0%) for it not to be disbelieved.

The other sort of comparison, between two sample means, can also be illustrated by an example. Suppose the performance of two estimators were being compared. The results of their recent estimates were different but could the difference have arisen by chance? The first estimator's errors were those in the previous example. The second estimator's errors were 28%, 31%, 21%, 25%, 6%, 11%, 19%, 39%, 49%. Their arithmetic mean is 25.4% and their standard deviation is 12.58%.

This time the t test is based on the standard error of the difference between the means of two samples drawn from the same population.

The estimate of the population standard deviation can this time be based on both samples because the null hypothesis is that they both come from the same population. It is calculated, as described in Section 5.1.2, from the squares of the separate standard deviations each multiplied by its own sample size. These two are added together as follows

$$10 \times 12.01^2 + 9 \times 12.58^2 = 2866.71$$

and divided by the total of the degrees of freedom for the two samples, i.e. $9 + 8 = 17$.

The estimated population standard deviation is the square root of the quotient,

$$\text{i.e.} \quad \sqrt{\left(\frac{2866.71}{17}\right)} = 13.0$$

The standard error of the difference between the sample means is the sum of the squares of the two individual standard errors. This will be slightly different in the case of the first sample from what was found in the previous example because the estimate of the population standard deviation is now improved by the inclusion of the second sample. The two standard errors are respectively

$$\sqrt{\left(\frac{13.0^2}{10}\right)} \quad \text{and} \quad \sqrt{\left(\frac{13.0^2}{9}\right)}$$

The square root of the sum of their squares is

$$\sqrt{\left(\frac{13.0^2}{10} + \frac{13.0^2}{9}\right)} = 13.0 \sqrt{\left(\frac{1}{10} + \frac{1}{9}\right)} = 5.97$$

Thus the difference between the sample means is 5.0 and the standard error of the difference is 5.97. Therefore the difference is 0.84 standard errors from zero. This can be looked up in the table of t against 17 degrees of freedom. The probability, double-tailed, that such a value or a greater one could have occurred by chance is more than 40%.

The conclusion must be that the results do not provide good evidence that there is any real difference in the performances of the two estimators.

With the sizes of samples usually available in the building field, reasoning of the sort just employed is not likely to mislead. However, if samples are very large, calculations like those above will show that even very small differences in sample means are unlikely to have occurred by chance. This is a common source of misunderstanding.

The term usually used to describe an effect shown to be unlikely to have occurred by chance is 'statistically significant', and it is this that causes trouble. The levels of probability usually looked upon as the criteria to determine whether the effect is statistically significant or not are 5% and 1%. If the probability that the effect could have occurred by chance is calculated to be less than 5% the effect is described as statistically significant at the 5% level. Sometimes it is said to be 95% significant. There is even a convention that an effect which is statistically significant at the 1% level is described as highly significant.

The distinction which must be drawn if the term 'significant' is used is between, on the one hand, statistical significance, which is affected by the amount of the difference, population variability and sample size; and on the other hand, practical significance, which is concerned only with the amount of the difference. A difference between two means from large samples can be statistically significant even though the difference is too small to be of practical importance.

This distinction causes no difficulty to statisticians so they often forget to point it out when reporting an analysis. In fact the distinction between statistical and practical significance has been used, most unjustly, to imply the impracticality of statistical methods. Again, an appreciation of statistical methods by those making decisions based on statistical analyses would be a great advantage.

The difficulty can be circumvented, given at least an appreciation of probabilities, by avoiding the use of predetermined thresholds of significance. It is better to calculate the probability of chance occurrence and treat each problem on its merits. The probability can usually be incorporated into the decision making.

5.3.2 Single- or double-tailed probabilities

In the above example, double-tailed probabilities were appropriate because the differences being examined could have occurred in either direction and both directions were of interest.

To help decide whether single- or double-tailed probabilities are appropriate, it is useful to ask, 'Did I know beforehand whether the deviations in which I would be interested would be positive or negative?' If so, then single-tailed probabilities are required. If not, then double-tailed probabilities are appropriate. A picture of a distribution with tails shaded, as in Figs. 5.2 (a), (b), (c), should be kept in mind.

78 *Statistical Methods for Building Price Data*

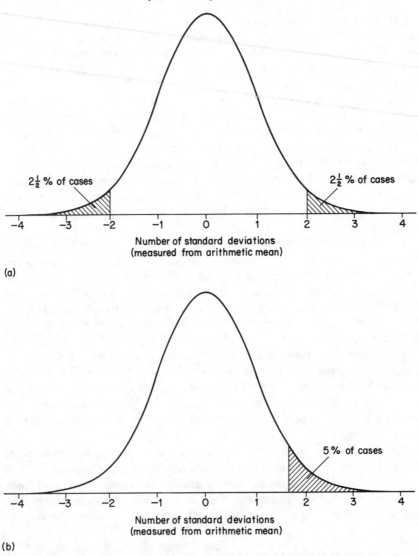

Fig. 5.2 Approximate proportions of area under the normal curve.

5.3.3 *The null hypothesis and limits of difference*

The null hypothesis is usually that two samples came from the same population and that the apparent difference between them is due to chance. If tests show that the probability is low that such a difference could have occurred by chance, the null hypothesis is rejected for the purpose of making a decision. It is often loosely said that the hypothesis is disproved. The use of the word

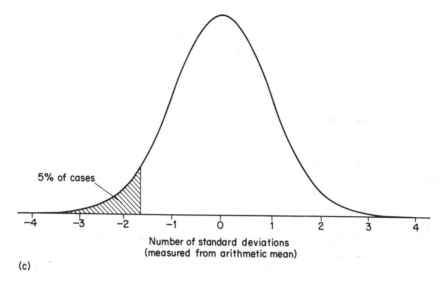
(c)

'disproved' is unlikely to be too misleading, but the opposite may be. It must not be thought that such tests can prove the hypothesis to be true.

The best that can be done is to say that, if the samples come from different populations, the true difference between them is probably less than a stated amount; or rather that there is a stated, low probability that the true difference is more than a certain amount.

This is done by first calculating the standard error of the difference between the parameters of the two samples. The usual parameter of interest is the arithmetic mean or the median. The calculation of the standard error of the difference between two arithmetic means is described in the example in Section 5.3.1. The standard error of the median is 1.25 times as great.

The maximum and minimum likely differences between the parameters are given by the confidence limits at the chosen level of probability, typically 0.05. Single-tailed probabilities are appropriate if either the minimum or maximum differences are of no interest.

In the example of the two estimators in which it was concluded that the difference could easily have occurred by chance, it would have been useful to add that the true difference was unlikely to be greater or less than stated amounts. These limits are obtained from the t-table, entered with 17 degrees of freedom and a probability of 0.05 (double-tailed). They are ±2.11 standard errors from the observed difference. The standard error of the difference of means was calculated as 5.97 so the required likely maximum and minimum for the difference is ±12.6 from the observed difference. The observed difference was 5.0, so the 95% confidence limits are −7.6% and +17.6%. It is because this range includes zero that it is concluded that the difference is not statistically significant.

Notice that although the two samples are not hypothesized to come from the same population, it is assumed that the two standard deviations can be pooled to estimate the population standard deviation. Thus, if the standard deviations are very different the confidence limits should be widened a little to allow for this. This is an approximation but it will serve for these purposes.

5.3.4 Non-parametric tests

The t test assumes that the populations from which the data come are normally distributed; but, in fact, it is not likely to be misleading even when the data are far from normally distributed. There are some tests which make no assumptions about the shape of the parent population. These are called non-parametric tests. They are mostly based on ranks rather than the actual values.

The merit of non-parametric tests lies not only in their applicability to a wide variety of data when nothing is known about the parent population; they are also less sensitive to a wildly wrong item of data. This is illustrated in the next example where the most generally useful of the non-parametric tests is applied to the performances of the two estimators, which were compared by the t test in Section 5.3.1.

5.3.5 The Mann–Whitney test

This test uses the ranks of the data to compare two samples by assessing the probability that samples as different or more different could have been randomly selected from the same population.

The first step is to rank the data for the two samples combined. This is

Table 5.2

Estimator A		Estimator B	
Data	Ranks	Data	Ranks
3%	1	6%	3
5%	2	11%	5
8%	4	19%	$7\frac{1}{2}$
14%	6	21%	9
19%	$7\frac{1}{2}$	25%	$10\frac{1}{2}$
25%	$10\frac{1}{2}$	28%	$12\frac{1}{2}$
28%	$12\frac{1}{2}$	31%	$15\frac{1}{2}$
30%	14	39%	17
31%	$15\frac{1}{2}$	49%	19
41%	18		
Rank sum 91		Rank sum 99	

quickly done when they are already separately ranked. The eye can then easily run down the two columns, noting against each item its rank in the combined sample. The ranks are then totalled for each sample, as in Table 5.2.

Straight away a comparison is available if the average ranks are calculated. These are 9.1 and 11.0; but this is not the basis of the Mann–Whitney test.

The test is based on setting down all possible arrangements of ranks and counting those which are the same as or are more favourable to sample A than the one which actually occurred. For small samples this is not too difficult and tables are available which give the probabilities corresponding to certain rank sums. For samples larger than 3 it is possible to use the table of the normal curve to evaluate the probability of a value of a rank sum.

The rank sum has to be converted to a deviation from the mean of its distribution and the deviation divided by its standard deviation. This expected mean and standard deviation are calculated as follows.

If the sample sizes are m and n, the mean of the distribution of the rank sum for sample A is

$$\frac{m}{2}(m+n+1)$$

and the standard deviation is

$$\sqrt{\left[\frac{mn}{12}(m+n+1)\right]}$$

In this example $m = 10$ and $n = 9$ so the expected mean for A is 100.0 and the standard deviation 12.25. The deviation from expectation of A's rank sum is $100.0 - 91$ but, because the rank sum can only be an integer whereas the expectation can be fractional, on average the deviation will be overestimated by $\frac{1}{2}$ so it must be reduced by this amount. This is called a continuity correction.

The deviation of 8.5 divided by the standard deviation is 0.69 and Table A gives the proportion of the normal distribution beyond 0.69 standard deviations from the mean in both directions as 49%.

Thus a rank sum as favourable to estimator A or more so could occur by chance with a probability of 0.49.

This result can be compared with the t-test which gave the probability of such a difference between the means or a greater one as 0.40. Exact agreement cannot be expected when the tests are so differently based. In neither case would it be concluded that estimator A was proved to be more accurate than B.

There are other non-parametric tests but the Mann–Whitney test is simple and appropriate to building price data.

In assessing the usefulness of any non-parametric test it is important to note that, even if the highest value had been very much higher than it was, the result would have been the same, whereas the t test would have been affected. The relevance of this quality to the analysis of building price data does not need stressing. It is more important than the usual reason for preferring a non-parametric test, the lack of dependence on knowing the form of the population distribution, because in building price data the field is narrow enough for the shapes of the likely distributions to be approximately known.

6 Data appreciation

6.1 Appraising data

So far, in the interests of showing as soon as possible how statistical methods provide a better way of viewing the behaviour of building price data, the practical aspects of handling raw data and making simple appraisals have been largely ignored. This chapter remedies this and explains to the reader who, the author hopes, is now convinced of their value, procedures for data handling which will inculcate good habits of data appreciation.

When the procedures are thoroughly familiar through frequent use the reader will have acquired the valuable ability to make rapid and consistent judgements from data without the need for unnecessarily elaborate analysis. Even when not using the procedures formally, but merely running an eye over a group of figures, some useful features will be discerned which would previously not have been noticed.

6.2 Arrangement of raw data

When data conforming to a chosen description have been selected, the first requirement is to inspect them. Summarizing methods have to be used to aid comprehension of the characteristics of the data and to provide simple measures to compare one set of data with others. Measures such as arithmetic mean, median, range, standard deviation, coefficient of variation and semi-interquartile range can be set beside those obtained from other data to provide an unbiased comparison. The word 'parameters', used for measures which help to describe a set of data, is derived from *para* = 'beside', and *meter* = 'measure'.

Bearing in mind the brain's ability to see pictures in the fire, spurious patterns may be discerned in well-arranged data which are, in fact, the result of chance. On the other hand, it is important to reveal all genuine patterns. The following advice on arranging data is designed to reveal their features. Statistical methods are explained in the rest of the book which help to distinguish genuine patterns from those which have occurred by chance.

The eye and brain have difficulty in extracting meaning from a set of numbers when they have more than two important digits each. When presenting data or making an initial appraisal, cutting them down to two significant digits should always be considered. The following example illustrates the value of this. The same set of 5 items are expressed to 4, 2, and 1 decimal places.

1.3584	1.36	1.4
1.9238	1.92	1.9
1.4142	1.41	1.4
1.3928	1.39	1.4
1.2871	1.29	1.3

If the only important feature of the data is that one of the 5 items is clearly greater than the other 4, the first set is unnecessarily detailed and does not quickly display its message. The third column is adequate for this simple comparison; but for most purposes, including the calculation of parameters, the second set gives the proper amount of detail. Although it has three significant digits, only two are important because when the figures before the decimal points are the same the eye takes them in altogether and is not thereafter distracted by them.

Even where the first digits are not identical, there can be the same effect if there are only two or three different figures and they are arranged in ascending or descending order. For example, the following numbers:

34.6
38.1
41.9
35.2
37.7
43.2

are readily comprehended when rearranged.

34.6
35.2
37.7
38.1
41.9
43.2

Vertical arrangements are easier to read than horizontal because the first digits can be more easily inspected separately from the others without the need for mental leapfrog. Therefore, in a two-way table, the figures most often or importantly to be compared should be under one another rather than across.

6.2.1 Frequency tables

Arranging data in ascending or descending order within their sets not only aids appreciation but improves accuracy when data have to be transcribed. If they are input to a computer or calculator with a printing or display device, checking for correct input is itself an error-prone procedure. Accuracy is greatly improved by limitation of the maximum error when there is the additional check that the data are in the correct order of size. The computer program to do this is very simple. Errors which do not affect the sequence are likely to be comparatively small.

Although complete sequencing is desirable, it is a rather laborious procedure so a half-way stage is useful. This is achieved by arranging the data in groups, each group containing data falling within a stated range.

To avoid ambiguities in defining the upper end of each range, only the lower end need be stated followed by a dash indicating 'up to but not including the bottom of the next range'.

In Table 6.1 the data are first recorded in the order in which they were collected and then allocated one by one to appropriate groups.

Table 6.1

1.36	1.92	1.41	1.39	1.29
1.60	2.08	1.07	1.58	1.64
1.77	1.87	2.03	1.97	2.21
1.51	1.51	1.77	1.71	1.30
1.76	2.18	1.24	1.96	1.15

Group	Data
1.00–	1.07, 1.15
1.20–	1.36, 1.24, 1.39, 1.29, 1.30
1.40–	1.51, 1.51, 1.41, 1.58
1.60–	1.60, 1.77, 1.76, 1.77, 1.71, 1.64
1.80–	1.92, 1.87, 1.97, 1.96
2.00–	2.08, 2.18, 2.03
2.20–	2.21
2.40–	

The last empty group is not essential but is a clear way of showing that there is none more than 2.40. For the sake of brevity the last group can be 2.20–2.40 or 2.20–2.39. '2.20 and over' fails to convey the information that the largest number is less than 2.40. It could be very large indeed, in which case the loss of information would be important. With more data than this it would be useful to record on the left the number of items in each group.

Until experience has been built up, it is not easy to choose a cell width. Obviously it should be a convenient unit but it is impossible to formulate a rule which always produces a good picture of the distribution. If a rough guide is needed then a convenient figure with a range of half to one standard deviation is often the best for numbers of data fewer than 50. For more data, narrower cell widths can be used.

With these data there is no obvious ambiguity in defining the upper end of each class as 1.19, 1.39, 1.59, etc., because they are all recorded to two places of decimal so the problem of where to classify, for instance, 1.598 does not arise. However, the problem is present but hidden because 1.60 may have been the result of rounding 1.598 and ought to have been allocated to the group 'below 1.60–'.

Whenever a frequency table is constructed or used the question 'what are the lowest and highest numbers that could have found their way into each group?' should be asked. In the above example the answer in the case of 1.40– would be '1.395 000 . . . and 1.594 999 . . .' The mid-point of this range is 1.495 and not 1.500 as might be first thought.

The mid-points are used in the next section and the end-points are used later when the frequencies are expressed cumulatively and smoothed. Although the error in using 1.500 instead of 1.495 is trivial, with very large amounts of data it could matter. It is worth knowing about because it is an example of errors all being in the same direction. This is bias. It has to be taken more seriously than random errors, which tend to cancel out.

If there are too many data to record as actual numbers, a distribution of them can be compiled by recording a tally mark instead of the number. Blocks of five are usual, with the fifth being a diagonal stroke through the previous four.

```
1.00–    11
1.20–    1111
1.40–    1111
1.60–    11111
1.80–    1111
2.00–    111
2.20–    1
2.40–
```

6.2.2 Histograms

If the tally marks are recorded evenly without making blocks of five, the space occupied by the tally marks is proportional to their number. If they are turned sideways and their outline drawn round them the result is a *histogram*. The tally marks in Fig. 6.1(a) are not necessary and are omitted in Fig. 6.1(b).

In fact it happens that these data produce a smoother histogram if a class interval of 0.25 is used instead of 0.20. The reader may wish to construct it as an exercise.

(a)

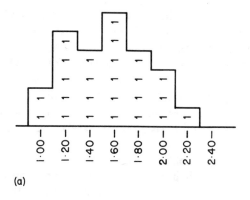

(b)

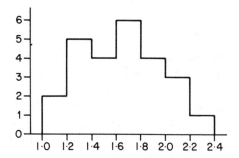

(c)

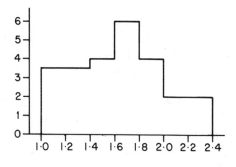

Fig. 6.1 (a) and (b) Forming a histogram from tally marks. **(c)** Combining cells.

It is not strictly necessary for all the cell widths to be the same in frequency tables and histograms. In frequency tables unequal cell widths can be misleading and are best avoided, but in histograms they are quite common and do not give a false impression. This is because the height of the histogram can be adjusted to accommodate any cell width. The principle which must be followed is that the area under any part of the histogram must be proportional to the frequency represented. To preserve this when the cell width is varied, the height of the affected parts of the histogram is proportionately changed to keep the area the same.

An example based on Fig. 6.1(b) will make this clear. If it is decided to combine the first two cells 1.0– and 1.2– their total frequency of 7 items must be shared equally between them. Thus the height of the combined cell is $3\frac{1}{2}$ as in Fig. 6.1(c), where the same rule has been applied also to the two right-hand cells.

6.2.3 Calculations from frequency tables

It is better not to use a frequency distribution for the purpose of calculation, but when very large amounts of data are to be processed and they are already in frequency table form, it may be desired to avoid using the basic data. To use a frequency distribution for calculation an assumption must be made about the distribution of the data within each class. The convention is to treat all data as though they lie at the centres of their classes. This assumes that the average of the data in each case is at the centre of the class.

Near the middle of the whole distribution this is a fair approximation because the data are likely to be distributed fairly evenly through the class, but elsewhere it is obviously not so because data tend to be concentrated more towards the side of the class nearer to the centre of the whole distribution. It would be better to guess the position of the average of the data in each class. The convention is fairly harmless because it has little effect on the average but its almost universal use is an interesting example of the common preference for a rigid rule rather than the use of judgement even when it is known that the rule is wrong and that judgement would be quite good and usually give a better answer than that given by the rule.

Whatever it is decided to assume about the position of the average of the data in each class, it should be recorded against the class and all calculations made with the assumption that all the data in the class had that value. Multiplying this value by the frequency in the class gives an estimate of the total of the values in the class. Adding these totals estimates the grand total of all values. This total is then divided by the total frequency to estimate the overall average.

Most calculating machines other than the most basic will accumulate the products of value and frequency without recording them individually. Some require the products to be added into a memory but most do not.

Data appreciation 89

The sum of squares of the data in a frequency distribution can be estimated in the same way. The square of the assumed value for each class is multiplied by the frequency and these products accumulated to estimate the total sum of squares. From this, together with the number of items and the estimated total of the values, the standard deviation can be estimated.

This estimate of the standard deviation is biased and must be adjusted by Shepherd's correction. If the class interval is h, then $h^2/12$ has to be subtracted before the final square root is taken to give the standard deviation.

6.3 Measures of location

Having considered the distribution of the data and the likely shape of the parent population, the next step is to calculate parameters which give a full enough picture of the data (or their parent population) not to need to display them again.

Parameters to measure the dispersion of the data have already been fully discussed in Chapter 2 and reference has been made to the need for measures of size, generally called averages. Statisticians prefer to describe them as measures of location.

'Average' is a term which covers all measures of location. It simply means 'true value' and is an attempt to sum up all the data in a single figure. It is a pity that so much attention is paid to it without taking into account the dispersion of the data.

6.3.1 Arithmetic mean

The most commonly used average is of course the total divided by the number of items – the arithmetic mean. In fact, when the term 'average' is used without further description it invariably means the arithmetic mean.

It is too much to expect a single figure to describe a distribution and it is obvious that the similarity of the arithmetic means of the three distributions in Fig. 6.2 conveys little about the data upon which they are based.

Possibly more misleading are the differences between the arithmetic means of the four samples in Table 6.2. All could easily have come from the same population if in the second a value from its lower tail had, by chance, entered the sample, and in the third a value from its upper tail. The high value in the fourth sample could be an error.

However, there are some uses for which the arithmetic mean gives all the information needed for the purposes for which it is calculated. In these the arithmetic mean is to be multiplied by the appropriate number of items to estimate the total. For example, if it is required to estimate the total cost of upgrading all public housing in a region by studying a random sample of houses, the arithmetic mean is the appropriate parameter because it simply has to be multiplied by the number of houses in the region to give an estimate

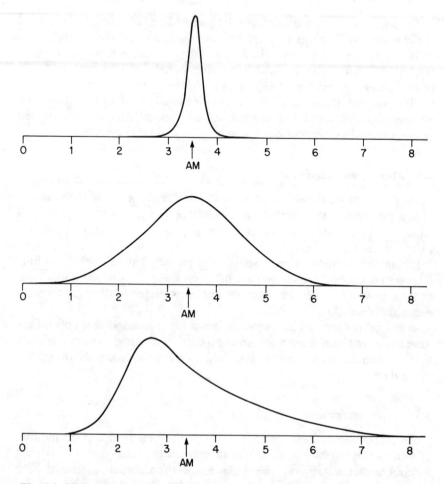

Fig. 6.2 Three distributions with the same arithmetic mean (AM).

Table 6.2

	4	1	4	4
	5	5	5	5
	5	5	5	5
	6	6	6	6
	7	7	10	20
Arithmetic means	5.4	4.8	6.0	8.0

of the total cost. However, if the shape and standard deviation of the parent population are known and the likelihood exists of very large occasional errors in the sample costs, it could be argued that using a measure to locate the distribution which is insensitive to the effects of a few gross errors and deducing from this the arithmetic mean would provide a better way of estimating the total than would calculating the arithmetic mean directly.

6.3.2 Trimmed mean

Sensitivity to extreme values is an important disadvantage of the arithmetic mean for users of building data. One way of partially overcoming it is to make a practice of omitting the highest and lowest portions of the sample from the calculation of the arithmetic mean. It is important immediately to stress that this must not be done when calculating a measure of dispersion.

For calculation of the arithmetic mean a reasonable trim is the omission of the top and bottom quarters of the sample distribution. This fits well with the use of the quartiles of the whole distribution to assess its dispersion and skewness. To some this may seem too large a trim if wild data are uncommon and samples large. They may prefer to use 10%.

It is best not to try to refine the calculation of the trimmed mean by omitting fractions of items. The number of items should be rounded to the nearest integer. In the example in Section 6.3.1 (Table 6.2) a 25% trim would remove one item (i.e. $1\frac{1}{4}$ items rounded to the nearest whole item) from each end of each sample leaving the same three values in each case. The trimmed means would therefore all be

$$\frac{5+5+6}{3} = 5\frac{1}{3}$$

It is sometimes objected that the trimmed mean throws away some of the data and so reduces the sample size. In fact the sample size remains unchanged. The data omitted from the calculation make their presence felt by deciding which items of data remain. The most convincing argument in favour of the trimmed mean, especially useful for persuading those reluctant to make a trim larger than 5% or 10%, is that the median is a 50% trimmed mean.

6.3.3 Calculating the arithmetic mean

When calculating an arithmetic mean, trimmed or otherwise, the commonest cause of serious error is the accidental omission of an item from the total followed by the use of the correct number of items to divide into it. Sometimes the total is correct but the number of items is wrongly counted.

Some calculating machines count the number of items which have been entered and avoid this error by ensuring that the number corresponds to the total. It is better for an item to be omitted from both than from just one of them. This should also be borne in mind when writing programs for a programmable calculator or a computer.

The item count is usually a by-product of the automatic calculation of the standard deviation, so it is often best to use this facility to calculate the arithmetic mean even when the standard deviation is not required.

To avoid dealing with large numbers it is sometimes helpful to subtract a constant from all items and add it back at the end. Since the advent of calculators and computers this device is less commonly used than it once was but if, for instance, all the data were between 2000 and 3000 it would ease the keyboard task to subtract 2000.

If a constant is subtracted its effect on other parameters must be considered. In most cases the effect, if any, is obvious. Measures of location are reduced by the amount of the constant subtracted and measures of dispersion are unaffected.

6.3.4 Geometric mean

An advantage of the arithmetic mean over the median is that it is calculated without the need to rearrange the data. This is also true of the geometric mean provided that either a calculator with a log button or a computer is used.

The calculator or computer is needed because the arithmetic mean of the logarithms of the data has to be calculated. The geometric mean is the antilog of this. It takes only a little longer to calculate than the arithmetic mean because it involves only one extra keystroke after each item is entered and then one extra at the end. The base of the logs does not matter but, of course, it must be the same throughout the calculation.

The application of the geometric mean to price data was described in Chapter 4 in connection with transformations. It gives a useful measure of location for data skewed to the right and is less affected by wildly high values than the arithmetic mean. It is more affected by wildly low values but these seldom occur in building price data.

Logarithms convert multiplication into addition, and taking roots into division. Therefore the sum of the logs of two numbers is the log of their product. Half the sum of the logs of two numbers is the log of their square root. Thus the geometric mean of two numbers is the square root of their product; of three numbers it is the cube root of their product. For more numbers it is hardly a practical way of calculating but the method is, in fact, the way in which geometric mean is defined. It is the nth root of the product of n numbers.

6.3.5 Median

Reference has already been made to the median as the middle value when the data are ranked in ascending or descending order. If there is an even number of data there are two middle values. The median is then midway between them and equal to their arithmetic mean.

Even if the data are not completely ranked, the median is easily found from a frequency distribution which records the actual data and not just their frequencies. There is an example of this in Section 6.2.1 (Table 6.1). The cell containing the middle value can be readily identified and only the items in that cell need be completely ranked.

The median is an excellent way of expressing the average of building price data because it is almost completely unaffected by the sort of wild data usually encountered.

6.3.6 Mode

The mode is meant to be the commonest value. In most samples this is vague because values seldom coincide. If all are different except for one pair of equal items, it is useless to call the value of the coincident pair the mode. It provides a very poor measure of location.

The only sensible way of obtaining a mode is from a frequency distribution when the cell with the greatest frequency can be said to contain the mode. The actual value of the mode can be taken as the centre of the cell but it is affected by the choice of cell boundaries.

The sensitivity of the mode to the choice of cell boundaries, as well as the way in which a distribution can change shape, is illustrated by manipulation of the distribution in Table 6.1.

As it stands the cell with the greatest frequency has a lower boundary of 1.60, so the mode could be taken as the centre of it at 1.70. However, if the cell interval had been chosen to be 0.25 instead of 0.20 the distribution would be as in Table 6.3.

Table 6.3

Group	Data
1.00–	1.07, 1.24, 1.15
1.25–	1.36, 1.41, 1.39, 1.29, 1.30
1.50–	1.60, 1.51, 1.51, 1.58, 1.71, 1.64
1.75–	1.77, 1.76, 1.92, 1.87, 1.77, 1.97, 1.96
2.00–	2.08, 2.18, 2.03, 2.21
2.25–	

The cell with the greatest frequency now has a lower boundary of 1.75. Its centre is 1.87.

If the class interval is restored to 0.20 but the boundaries shifted by 0.10 the distribution is as in Table 6.4.

Table 6.4

Group	Data
0.90–	1.07
1.10–	1.24, 1.29, 1.15
1.30–	1.36, 1.41, 1.39, 1.30
1.50–	1.60, 1.51, 1.51, 1.58, 1.64
1.70–	1.77, 1.76, 1.87, 1.77, 1.71
1.90–	1.92, 2.08, 2.03, 1.97, 1.96
2.10–	2.18, 2.21
2.30–	

There are now three cells with a maximum frequency. The centre of them is 1.80.

Thus the mode as deduced from frequency distributions has been 1.70, 1.87 and 1.80. Smoothing the distributions usually provides more stability and unless there is time to do this it is best not to use the mode. However, smoothing the histograms would still give different results and much would depend on opinion. Different people would smooth differently and the reader may wish to use these examples to see whether, like the author, he still gets uncomfortably differing results.

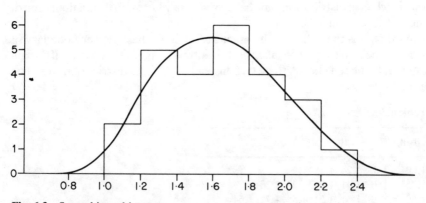

Fig. 6.3 Smoothing a histogram.

The simplest satisfactory way of smoothing is to try to make the area under the curve the same as that under the histogram at all parts of the scale as far as this can be done without sacrificing smoothness. Figure 6.3 is one attempt at this based on the histogram in Fig. 6.2. The mode of this curve is 1.6.

The mode is unaffected by wild data so it has a superficial attraction for users of building price data even though it is erratic and difficult to obtain. Also, it has a few genuine applications in the building field which may be encountered in association with building price data. It is needed to convey such information as the commonest number of bedrooms in houses and the most likely family size. Of course, it would be better to give more information about the distribution than just the mode.

It is possible for a distribution to have two quite distinct peaks, whatever class boundaries are used. Such distributions are described as *bimodal*. The effect is usually no more than a quirk of the particular set of data and only indicates a true characteristic of the factor under consideration if the data are a mixture of two very distinct distributions with different modes. If this is the case the two sets of data should be separated and analysed separately.

6.3.7 Weighted mean

When calculating an arithmetic mean or geometric mean it may be wrong to give every item of data equal weight. When cost per unit area data from previous analyses are used to build up an overall estimate of cost per unit area it would obviously be wrong, for instance, simply to average the figures for each floor regardless of area when the areas differ markedly. The cost per unit area assumed for the ground floor of a building must be multiplied by its floor area before being added to the corresponding products for other floors. This is no more than the obvious procedure of converting each cost per unit area to total cost, totalling these and dividing the grand total by total floor area. Similarly, if costs per unit floor area from several buildings are to be weighted according to their size, the size measure should be floor area.

Before deciding whether and how to weight data, the objective must be carefully considered in the light of the use to which the results will be put. If the data are weighted by building size, very large buildings will dominate the arithmetic means and this may or may not be desirable. If large buildings have lower costs per unit area, the result of calculating a mean weighted for size would be a cost per unit area too low for application to small buildings.

The best solution would be to establish the relationship of cost per unit area to size and to adjust the data to a common building size before averaging. Before use the average can be adjusted to the desired building size.

If this solution is not available, perhaps the data can be divided into a few size groups and separate results calculated. The difficulty here is that the difference between the size groups may be due to chance. However, this is effectively the method used when the buildings selected from a data bank are chosen on the basis of size. If only small and medium-sized buildings are appropriate to the task then large buildings will not be selected.

If some weighting is felt to be necessary but not the full amount, it is likely that the square root of the area or of the total cost will have the desired effect. Logarithms to any base reduce the effect even more than square roots, as noted in the descriptions of the geometric mean and of transformations.

A more controversial procedure is to weight data in accordance with their relevance or reliability. This is recommended in the section on combining data and the use of judgement in Section 3.6.1.

The introduction of a subjective element must be remembered when the mean is used and it is important that the effect of the item on the answer is not allowed to influence the weight given. An item of data must never be given a low weight simply because it is too high or too low to be believed. The weight should be allocated before the value of the variable of interest is known.

Whatever method of weighting is decided upon, the weighted mean is calculated in the obvious way by multiplying the weight for each item by the value of the variable, totalling these products and dividing this total by the sum of the weights.

6.3.8 Weighted median

This measure of location has the same advantages as the median when wild data are present in the sample and allows weights to be used. The best way of explaining its calculation is by example. The example in Table 6.5, for five buildings, shows first that the weighted median is close to the weighted mean when the data are fairly symmetrical, and second the effect of wild data.

Table 6.5

Cost per unit area (in ascending order)	Floor area	Weight $= \sqrt{\text{area}}$ (W)	Cumulative weight (C)	Centred cumulative $(C - \tfrac{1}{2}W)$
73	1348	36.7	36.7	18.4
89	1921	43.8	80.5	58.6
132	853	29.2	109.7	95.1
144	1114	33.4	143.1	126.4
201	3649	60.4	203.5	173.3
		203.5		

Half the total of the weights is 101.75. The centred weights show that the median lies between 132 and 144. To find whereabouts in the interval between 132 and 144 the median lies is a proportioning exercise using the centred weights.

The required proportion of the interval 132 to 144 is

$$\frac{101.75 - 95.1}{126.4 - 95.1} = 0.2125$$

This proportion is applied to the costs per unit area to give the median:

$$132 + 0.2125(144 - 132) = 135$$

For comparison the weighted mean is calculated as in Table 6.6.

Table 6.6

Cost	Weight	Cost × weight
73	36.7	2 679
89	43.8	3 898
132	29.2	3 854
144	33.4	4 810
201	60.4	12 140
	203.5	27 381

The weighted mean is 27 381 ÷ 203.5 = 135.

The effect of wild data is shown by imagining the greatest cost to be increased from 201 to 301. The weighted mean rises from 135 to 164 whereas the weighted median remains unchanged.

6.3.9 Harmonic mean

The harmonic mean is more a curiosity than a useful measure of location. It is calculated by taking the arithmetic mean of the reciprocals of the data and taking the reciprocal of the answer. It is difficult to find applications for it because in practice the calculation would be done from first principles.

If equal quantities of concrete were laid at different rates, the overall rate would be the harmonic mean of the rates. However, it would be satisfactory to calculate the total quantity of concrete and the total time.

6.4 Rapid appraisal

Careful calculations are necessary for thorough data analysis but it is impracticable to make a complete analysis of every set of data. There is not always time or effort available. A full analysis would include, as a minimum, a frequency distribution, measures of dispersion and location and a comparison of these with what was expected.

It is useful to be able to extract the maximum of unbiased information from data with the minimum of effort. The object would be to obtain an idea of the characteristics of the parent population without a full analysis. Practice is necessary to be able to do this and the reader is recommended to develop a standard set of procedures. The methods given here can be adapted to suit the facilities available.

Rapid appraisal methods can also be used to give a rough forecast of the results of a more thorough analysis. This helps to prevent gross errors and may indicate the appropriate type and depth of analysis. Sometimes, when the forecast turns out to have been a poor one, the reasons why simple methods were misleading gives a further insight into the characteristics of the population being studied.

It is not enough just to look at the data, because the eye tends to seek patterns. These are sometimes informative but they are so attractive that they distract from other more important features. Also it is difficult to avoid over-emphasizing the one or two wild figures which catch the eye.

The methods chosen will depend on how much data are being appraised and the amount of time it is reasonable to devote to the task. If this is only a few seconds, even the simplest of these procedures would take too long. On such occasions the habits formed by their frequent use will increase the amount of information which can be extracted by merely running an eye over the data.

In Table 6.7 the methods are classified according to the number of items and the time available. It is assumed that the data have already been made comparable with those in the data bank or price book by adjustment for inflation with a price index. It is also assumed that a data bank or price book exists which records expected averages and which can be annotated with expected coefficients of variation.

There is no need to be specific about the meaning of 'few' and 'more' in the headings but, if guidance is needed, about ten items or so would qualify as 'more' for many users.

Table 6.7 Rapid appraisal methods

Few items and little time	Few items and more time	More items and little time	More items and more time
Consider what coefficient of variation (c.v.) or standard deviation (s.d.) and average could be expected for the population which this sample represents			
Count the items in the sample			
Rearrange the data in ascending order. Obtain the median and range.	Rearrange the data in ascending order. Obtain the median and quartiles. Calculate the arithmetic mean.	Note the extremes and calculate the range. Average the extremes to help guess the median. Count the items greater than the guess and thus adjust the guess.	Make a frequency table. Obtain the medians and quartiles. Calculate the arithmetic mean and, if there is time, the standard deviation.

Table 6.7—*continued*

Few items and little time	Few items and more time	More items and little time	More items and more time
Roughly assess skewness by comparing:			Inspect the distribution for evidence of two populations and of skewness. Compare the median with the quartiles and the arithmetic mean to assess the skewness. If the s.d. was calculated, compare the number of items beyond one s.d. above the mean with the number beyond one s.d. below (expect 1/6 each for normality).
the median with the extremes.	the median with the quartiles and extremes, and with the arithmetic mean.	the median with the extremes.	
Estimate the population standard deviation by:			
(a) dividing the range by $\sqrt{n}$, or (b) using a calculator with s.d. facility. Use the $\sigma(n-1)$ button.	(a) calculating the s.d. of the sample and applying Bessel's correction, or (b) using a calculator with a $\sigma(n-1)$ button.	using Table B and the range.	multiplying the interquartile range by $\frac{3}{4}$; unless the s.d. of the sample is already calculated, in which case apply Bessel's correction to it. (The population s.d. may have been estimated directly using the $\sigma(n-1)$ formula) button on the calculator.

Estimate the standard error of the median as $\dfrac{1.25 \text{ (population s.d.)}}{\sqrt{n}}$

Estimate the population coefficient of variation by expressing the s.d. as a percentage of the arithmetic mean or median. If the median is used, slightly reduce the c.v. if the distribution is judged to be strongly skewed to the right.

Compare the median with what had been expected.
How many standard errors separate the observed and expected medians? This is *t*. Look it up in the *t* table (Table C) to find the probability that such a difference, or a greater one, could have been produced by chance.

Table 6.7—*continued*

If it is decided that the difference was probably due to chance, compare the standard deviation with what had been expected in the same way. Its standard error is (population s.d.)/$\sqrt{(2n)}$. If it is decided that the difference in medians was probably not due to chance, compare the coefficient of variation with what had been expected. Its standard error is (population c.v.)/$\sqrt{(2n)}$.

What can be learned from the comparisons of medians and s.d.s (or c.v.s)?
Reassess the expectations of the average and population c.v. if this seems indicated. Amend the data book or other record of expected population parameters.

6.4.1 Examples of rapid appraisal

To check data in a cost data bank, times to carry out the same site operation were measured. The sites were chosen at random from contracts larger than a certain value being carried out by a contractor at the date of the selection – work in progress. The population from which they came might ambiguously be described as 'all contracts larger than the chosen value undertaken by the contractor', but bias in selection would have to be considered. If the results of the analysis were to be applied to contracts as they were obtained, the distribution of construction durations would be biased, because a higher proportion of shorter contracts would pass through the books than would be present at any one time. As the data were sampled from work in progress it would be biased in comparison with the population to which the results would be applied. The restriction to high-value contracts would reduce the bias but it would still be present.

In this case it will be assumed that size of contract does not affect the site operation being examined. If this were not permissible the sample could be stratified on construction duration and weighted by the reciprocal of the average duration in each stratum. This could only be done with a sample at least as large as the second one used in these examples. Alternatively, using methods described later, the relationship between operation time and construction duration could be measured. This should be of more value than a single average.

For all the following examples, the existing data book record shows the 'standard' time, based on measurements made many years ago, to be 25 hours. No specific measure of dispersion is given but the times are said to vary between 20 and 30 hours 'under normal conditions'. This implies a symmetrical distribution with a median and arithmetic mean of 25 hours. Because abnormal conditions are common enough in building to affect as many as 20% of the data, 20 and 30 hours will be taken, arbitrarily, to embrace 80% of data. The reasonable assumption of a normal distribution will be made. Since 80%

Data appreciation

of a normal distribution is contained within ±1.3 standard deviations of the mean, 20 to 30 covers 2.6 standard deviations. The standard deviation can therefore be taken to be 3.8, making the coefficient of variation 15%. Notice that this calculation would have been the same even if the average had not been exactly mid-way between 20 and 30.

(a) Few items, little time
Consider seven items of data: 16, 22, 25, 15, 22, 26, 18.
  First they must be put in ascending order:

15
16
18
22
22
25
26

The median is 22 hours and the range is 11. The lowest is further from the median than is the highest but the evidence of skewness is not clear in such a small sample.
  The estimated population standard deviation is

$$11/\sqrt{7} = 4.2 \text{ (the exact calculation gives 4.3)}$$

The estimated standard error of the median is

$$1.25 \times 4.2/\sqrt{7} = 2.0$$

The median is 3 less than expected. This is 1.5 standard errors and not unlikely to have been due to chance. (This is obvious without using the t-table).
  Roughly blending the old and new data, a revised median of 24 hours would be reasonable.
  The coefficient of variation is about 20% and could be noted instead of the tentative unrecorded figure of 15%.

(b) Few items, more time
Consider the same seven items:

15
16
18
22
22
25
26

The median is 22 and the quartiles are $16\frac{1}{2}$ and $24\frac{1}{4}$.

The arithmetic mean is 20.6.

The evidence of skewness to the left should be regarded as only suggestive in such a small sample but could be noted for future confirmation if more data became available. It may be worth asking whether those on site take special speed-up action if the operation looks like taking longer than a certain time.

The standard deviation of the sample is 3.99 which, when multiplied by Bessel's correction, $\sqrt{(7/6)}$, becomes 4.3. This is the estimated population standard deviation. If by now the calculation is thoroughly understood so that no confusion arises, it can be made slightly easier by using $n - 1$ instead of n in the formula for standard deviation. If the calculator has a button marked to indicate the '$n - 1$ formula', this can be used to obtain the estimated population standard deviation directly.

The standard error of the median is

$$1.25 \times 4.3 \div \sqrt{7} = 2.0$$

The estimated population coefficient of variation is

$$4.3 \div 20.6 = 0.21 = 21\%.$$

The median is 3 less than expected and this is 1.5 standard errors.

The probability (double-tailed) of a value of t greater than 1.5, with 6 degrees of freedom is 0.2 so the difference of the median from the expected value could fairly easily (with a probability of 0.2) have occurred by chance.

The standard error of the standard deviation is $4.3/\sqrt{14} = 1.15$. As the observed and expected standard deviations differ by only 0.5 they are entirely compatible.

As no measure of dispersion was recorded in the data book, a coefficient of variation of 20% could be entered.

Thus the new data do not provide good evidence of a change in the time for carrying out the operation or that the standard time recorded in the data book is wrong. However, the recorded figure may have been crudely assessed. If so, it would be reasonable to blend the old and new averages using weights which reflect their relative soundness.

If the recorded average was thought to have been based on more observations than the new figure, it may be considered to be worth more weight. Assuming it was based on 20 observations but that they were made under less controlled conditions, their weight could be arbitrarily halved to produce a weight of 10. The new figure was based on a sample size of 7 so the blended median would be

$$\frac{10 \times 25 + 7 \times 22}{17} = 24 \text{ hours}$$

If the figure is going to be used as a multiplying factor, the recorded average would have been an arithmetic mean so it should be blended with the new arithmetic mean. The new standard time would then be

$$\frac{10 \times 25 + 7 \times 20.6}{17} = 23 \text{ hours}$$

(c) More items, little time

The following data are for 20 observed times instead of 7:

```
16  26   8  19
22  18  20  12
25  23  20  28
15  13  15  21
22  16  24  10
```

The shortest time is 8 hours and the longest 28 so the range is 20. (Run the eye down the columns, not across.)

The first guess at the median is $(8 + 28)/2 = 18$. Of the 20 items 11 are above this, so try 19. There are 10 above this and one equal, so there must be 9 below. Try 20. As there are 2 items equal to it this would give 8 above and 10 below. The nearest to equality is 10 above and 9 below, so the median is $19\frac{1}{2}$.

The median is a little above the mid-point of the range (the first guess). This could provisionally be interpreted as a suggestion of skewness to the left.

Table B shows that for a sample of 20 items the estimated population standard deviation is

$$\frac{\text{range}}{3.73} = 5.4$$

The standard error of the median is

$$\frac{1.25 \times 5.4}{\sqrt{20}} = 1.5$$

The estimated population coefficient of variation is roughly

$$\frac{5.4}{19.5} = 0.28 = 28\%$$

The median had been expected to be about 25 hours. The new estimate is 3.7 standard errors lower.

$$\frac{25 - 19.5}{1.5} = 3.7$$

For 19 degrees of freedom the probability of the difference being as great as or greater than this is about 0.001 so the difference is almost certainly not due to chance.

The standard error of the estimated coefficient of variation is about

$$\frac{28}{\sqrt{40}} = 4.4$$

The expected coefficient of variation was about 15% so the new estimate is 3.0 standard errors higher.

$$\frac{28 - 15}{4.4} = 3.0$$

This difference is almost certainly not due to chance and the reason is likely to be the poor estimate made from the information previously available. This new estimate (28%) is soundly based and should be recorded in the data book.

The new median could be recorded as $19\frac{1}{2}$ but, bearing in mind that its standard error is 1.5, it would be reasonable to acknowledge that the old figure was higher by entering a figure in the range $19\frac{1}{2}$ to $21\frac{1}{2}$ hours into the data book. The obvious entry would be a round 20 hours with a coefficient of variation of 30%, or a standard deviation of 6 hours.

(d)   More items, more time
Consider the same 20 items as in the previous example. The choice of cell width for a frequency table is not easy. The standard deviation is $5\frac{1}{2}$ so the rough guide of $\frac{1}{2}$ to 1 s.d. indicates a cell width between 3 and 5. Perhaps 5 is rather large but it shows the shape and is easy to use.

Group	Data
5–	8
10–	13, 12, 10
15–	16, 15, 18, 16, 15, 19
20–	22, 22, 23, 20, 20, 24, 21
25–	25, 26, 28
30–	

Obtaining the median is fortuitously easy because the first three cells contain half the items. The median must be the average of the largest number in the 15– cell and the lowest in the 20– cell. It is $19\frac{1}{2}$.

As before, the quartiles are 15 and $22\frac{1}{2}$.

The arithmetic mean is 18.65 and the estimate of the population s.d. is 5.46.

The median is nearer the upper quartile than the lower and a little greater than the arithmetic mean.

There are 3 items beyond one s.d. above the mean and 4 beyond one s.d. below (1/6 of the items, i.e. $3\frac{1}{3}$ would be expected in a normal distribution, so this is perfectly compatible).

The very slight skewness to the left in the sample could very easily have been due to chance. If slight truncation of the longest times is known to occur, this sample is consistent with it but must not be taken as evidence for it.

If the standard deviation had not already been calculated, the population s.d. could be estimated from the inter-quartile range. The IQR is $22\frac{1}{2} - 15 = 7\frac{1}{2}$. Multiplying this by 0.75 gives an estimate of 5.6 for the population standard deviation.

The standard error of the median is

$$\frac{1.25 \times 5.46}{\sqrt{20}} = 1.53$$

The estimated population coefficient of variation is 29% (mean/s.d.) and, as before, both this and the median are too different from the previously expected figures for the differences to be ascribed to chance. Using the same reasoning as before, the new entries in the data book could well be 20 hours for the median and a coefficient of variation of 30%.

6.5 Watching for changes

6.5.1 Control charts

Data are not always present for analysis together. Sometimes they flow in gradually and if it is important to detect changes in the average the methods of quality control may be applicable.

Estimators should keep records of their performance. A contractor's estimator might wish to monitor changes in the ratio of his estimate to the median tender. Perhaps he might prefer to compare his estimate with the lowest tender (excluding his own firm's) or he might devise some other measure of his performance.

Certainly the client's estimator would want to compare his estimates with the tenders and would probably choose the lowest or the accepted tender as representing his target. If he chose the accepted tender he would have to consider whether the acceptance would be based on closeness to his estimate and thus bias the comparison.

Whatever target is chosen, the comparison should be plotted on a simple graph, as in Fig. 6.4.

The graph becomes more informative if the long-run mean error and probable upper and lower limits obtained as follows, are drawn as horizontal lines.

Past data should be analysed to measure the arithmetic mean and the dispersion. The distribution is likely to be normal, remembering that an early name for the normal distribution was the error distribution, in which case it is only necessary to calculate the arithmetic mean and standard deviation of the past data to enable the percentage points to be obtained from a table of the

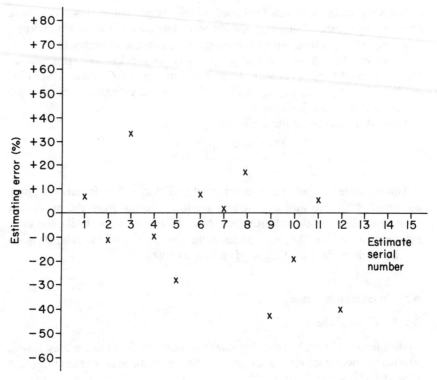

Fig. 6.4 Plotting estimating errors

normal distribution. If the distribution is slightly skew, an allowance for this can be made. If it is very skewed to the right the log or the square root of the data can be used throughout.

It is usual for two sets of limits to be drawn. If any data fall outside the outer pair it is taken to indicate that there has been a deterioration in performance. Data between the inner and outer pairs are considered to provide a warning that change may be occurring.

The outer limits are often called *action limits* and they should be drawn at the 99% or even 99.9% points. The inner, or warning, limits can be at 90% or 95% points.

Consider a record of percentage estimating errors with an arithmetic mean of +2 and a standard deviation of 20. If the warning limits are to be at the 95% (two-tail) points, they must be drawn at 1.96 standard deviations above and below the mean. The action limits would usually be at the 99% points and be drawn at 2.58 standard deviations above and below the mean.

The limits have been added to Fig. 6.5.

While the incoming data stayed within the limits, the estimating process would be described by quality control engineers as under control. If many

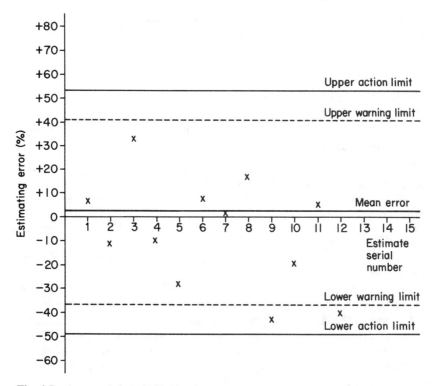

Fig. 6.5 A control chart for estimating errors.

more than 5% fell beyond the inner limits and more than 1% were beyond the outer limits the process would be said to be out of control and require action. It would be noted that a deterioration had occurred and reasons looked for. The second point below the lower warning limit would give cause for concern.

Although it would be unrealistic to use a standard deviation less than that measured in historic data, it is possible that action could be taken to adjust the mean error to be close to zero. For example, the estimating methods could be kept the same but a final adjustment made to the estimate equal and opposite to the previous mean error. This ought to shift the mean error to zero. Another way would be to adjust the data used to provide estimates.

If it is expected that the future mean error will be zero, or some other desirable figure, it is reasonable to draw all the control limits so that they are symmetrical about zero. If a change of estimating method or policy is made it may be necessary to draw the chart again basing it on new data.

Control charts were originally devised for monitoring quality in production processes and there may be similar applications in construction. The classic method is to take equal-sized random samples from production and plot the arithmetic means. The control limits are then equal to confidence limits

108 *Statistical Methods for Building Price Data*

calculated from the standard error of the mean. For some processes it is also of interest to use a control chart to monitor the ranges of the samples.

6.5.2 Cusum charts

Another technique designed to detect changes in data as they flow in is the cusum chart. This records cumulative differences from a target figure, so it is especially applicable to monitoring estimating percentage errors. The target in that case would be the expected mean percentage error, as in the control chart.

Table 6.8 Estimating errors. Target = +2%

Percentage errors	+7	−11	+33	−10	−28	+8	+2	+17	−43	−19	+6	−41
Differences from target	+5	−13	+31	−12	−30	+6	0	+15	−45	−21	+4	−43
Cumulative differences	+5	−8	+23	+11	−19	−13	−13	+2	−43	−64	−60	−103

As long as the data hover randomly around the target figure the cusum graph wanders about the horizontal, but as soon as there is a tendency for the data to be above the target the cusum graph begins to climb. A tendency for the data to be below the target is shown by a fall. Only the general direction of movement should be noted, not how high or low the level happens to be. The only useful indication is a sustained rise or fall.

Figure 6.6 is a cusum chart of the data in Table 6.8. The downward tendency after serial 8 indicates a tendency for the estimating error to be below target.

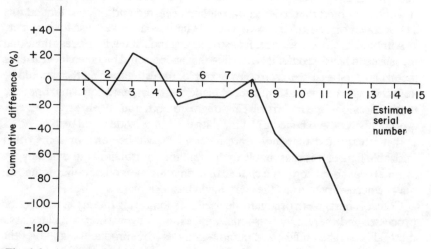

Fig. 6.6 A cusum chart.

Data appreciation 109

As the level of the cusum graph is of no importance, and to avoid running off the paper, it can be restarted at any new level. If it seems desirable the target can be changed to make the cusum chart horizontal. The monitoring process consists of watching for changes in slope.

The cusum chart seems a rather strange device but some users find it a simple and sensitive method for early detection of change. The information it gives can be compared with that from a control chart by comparing Figs 6.5 and 6.6, which are both based on the data in Table 6.8.

7 Tender patterns and bidding strategy

7.1 A model for tender patterns

Contrary to general belief, a typical distribution of competitive tenders for the same contract is almost symmetrical. In fact it is closely approximated by the normal distribution. There is a very slight skewness to the right but this is so small that it can for practical purposes be ignored.

On rare occasions there are enough (about 20) bids for a single contract to provide a good indication of shape, and then the characteristic indications of conformity with the normal distribution are seen. When single contracts with such large numbers of bids are not available, a way must be found to combine contracts even though they have different values.

In each contract the bids must be standardized. This can be done by expressing each bid as a percentage of the arithmetic mean of all the bids for the contract. Other standardizing figures could be used but, as all are to be combined, the arithmetic mean is satisfactory and convenient, especially for automatic processing.

The resulting percentages can be regarded as items of data in a single distribution. When more than 50 contracts are used, and preferably 100, the combined distribution is smooth enough to give a clear confirmation of a very slightly skewed normal distribution.

Conformity with the normal distribution with only negligible skewness is not what would be expected from most people's mental model of tendering. In this model the tenderers who most want to win the competition estimate carefully and bid low. Their bids can be expected to be close to each other and include the winner. Those less keen estimate less carefully and pitch their tenders higher. Their bids should be higher than the others and more scattered. Some of them would not be properly estimated but pitched high enough to ensure that the bidder would not win and intended only to avoid telling the client that he was not interested in tendering for the project.

The total effect of such a model would produce a distribution very strongly skewed to the right and the following would be a typical idealized pattern of bids conforming to it:

180 000
160 000
153 000
148 000
145 000
144 000

In fact, analysis of large numbers of tenders shows that a typical pattern would be more like the following, also idealized:

178 000
169 000
163 000
158 000
152 000
144 000

Conformity with the normal distribution implies a central tendency with random deviations from the centre, so a possible model is one in which bidders aim for much the same price initially and then add or subtract from it. The addition or subtraction could represent differences in keenness and estimating error.

Because the shape of the distribution is close to being normal it suggests that estimating error is the major cause of the deviations. It does not mean that bidders consciously aim for the centre of the list. They use methods which, on average, put them in the middle and the ones that they win are those where their downward errors are greatest.

This model, although untested, explains the results better and more simply than others. If varying keenness were the main influence it would seem likely that less keen bidders would add larger sums than keener ones would subtract. The very small amount of skewness indicates that this is a comparatively unimportant effect.

Some confirmation of a 'random error' model is that attempts to rank the bidders before opening tenders are only as successful as would be expected by chance.

If in some field it is found that the distribution of tenders is markedly skew it could be taken to imply that varying keenness has a more important effect than usual or their estimating errors are less.

When comparing the contributions to tender variability which may have been made by keenness and error, the way in which they combine by squares must be borne in mind. If the contributions were equal and the combined coefficient of variation were 7% they would be as follows:

112   *Statistical Methods for Building Price Data*

Cause of variability

	c.v.	c.v.^{2}
Keeness	5%	$24\frac{1}{2}$
Error	5%	$24\frac{1}{2}$
	7%	49

Correlation between them may increase their individual contributions but the argument is unaffected.

If error were more important, the results could be as follows:

	c.v.	c.v.^{2}
Keenness	$3\frac{1}{2}$%	13
Error	6%	36
	7%	49

So, saying that error makes a greater contribution than keenness is not saying that variability due to keenness is necessarily very small.

7.2 Non-serious tenders and cover prices

A tender may be submitted by a contractor who has been included in a restricted tender list and feels that he ought to bid even though he does not want the contract. Such a bid can be called a 'non-serious tender' and a special case of it is the 'cover price'.

A cover price is not arrived at by careful estimating but by asking another tenderer for a figure which is safely higher than his bid will be. Other non-serious tenders result from rough estimating and a very high mark-up. They have already been mentioned as an ingredient of the commonly held mental model of the tendering process.

The very slight skewness produced by these tenders may indicate that they are rarer than is generally believed or that rough methods of estimating are not so greatly inferior to careful methods that they increase the total variability by much.

In a sense these bids are genuine tenders because the contractor would do the work somehow if by any chance his high bid were accepted. The high bid, like any other high bid, reflects a low level of keenness. Such bidders do not want to be regarded as being reluctant to tender but neither do they want to get a reputation for high prices, because they feel that both would reduce their chance of being included in future tender lists. This probably helps to keep their tenders within the same normal distribution as the others.

7.3 Dispersion of tenders

In construction contracts in the UK and USA the combined coefficient of variation of competitive tenders for the same contract is between 5% and 9%

for most types of work. However, it is not constant. At times the average exceeds 10%. These times occur when, for instance, the rate of tender price inflation changes.

Some tenderers respond to changed conditions before others and some more violently than others, so for a few months the dispersion increases before returning to its previous value when contractors get used to the new rate of inflation. Notice that, as for the cusum chart, it is the rate of change which is influential, not the level.

Incidentally, the same upset in inflation rates causes average errors of clients' estimators to change similarly. The effect on their dispersion is less marked, probably because their dispersion is greater.

Differences in coefficients of variation in tender lists between types of building or types of work can be instructive. For example in the UK the variability in tenders for mechanical and electrical engineering contracts is greater than that for building contracts. This indicates larger errors in contractors' estimates of their costs. The reasons for this are no doubt the same as those which cause the larger estimating errors made by clients' estimators in mechanical and electrical engineering contracts than in building contracts.

A likely cause of differences in the variability of estimates is differing amounts and quality of information provided to contractors. This shows how the study of variability in tenders can be rewarding when tenders are massed, each having first been expressed as a percentage of the mean for the contract. There is also something to be learned from the coefficient of variation of the tenders in an individual contract when faced with a difficult decision on whether to accept the lowest or to call for tenders again in the hope of getting, by chance, a lower one.

If retendering is undertaken under the same conditions as before, it must be assumed that the average of the tenders will be about the same. Of course it may turn out to be different but the best guess is that it will be about the same.

If the coefficient of variation in the field of activity is known from previous analyses, the expected value of the amount by which the lowest tender is below the average can be obtained from Table B. Halving the expected range for the appropriate number of bids gives the amount as a multiple of the coefficient of variation. If the lowest tender is much above expectation it is reasonable to hope that, on retendering, the average will be about the same but the dispersion greater, so that a lower bid will be received.

This cannot be a precise exercise but, like other statistically based strategies, if consistently followed it will give better results in the long run than other ways of viewing the problem.

With great experience of analysing tender variability, it may even be possible to obtain indications of collusion if it occurs on a large enough scale. Nothing could be deduced from a few contracts but if consistent perturbations from the expected normal distribution of tenders occurred in a particular field of activity it would be sensible to look for reasons and the possibility of collusion should not be excluded. This is far from saying that evidence of collusion is provided; at most it can show where to investigate.

7.4 Bidding strategies

Some strategies offered to contractors depend on analyses of the contractor's performance against specified competitors. These are unlikely to be successful for the following reasons: specifying competitors greatly reduces the data upon which the strategy can be based; there is little chance that when applying the strategy it can be applied to more than one or two competitors in any one tender list; it is not always possible to discover who the competitors are before having to tender; an individual competitor may change his strategy thus rendering past data about him misleading; and lastly, such strategies seldom take account of keenness to win. This does not mean that special knowledge about one or more competitors is not useful, but such knowledge is best used to adjust a bid which has been constructed initially in a consistent way without using that knowledge.

Although treating competitors individually is unlikely to be a useful strategy, it is sometimes written about and it may be as well to correct a false impression given by some explanations. The fallacy was pointed out by R. M. Skitmore of Salford University in a private communication. It is as follows.

Past data collected by a bidder show that, in competition with a particular competitor A, the proportion of times when he beat A was $P(A)$. For other competitors the proportions were $P(B)$, $P(C)$, and so on. These proportions, or some adjusted version of them, are taken to be the probabilities of beating the competitors individually. The false conclusion is then drawn that the probability of beating all of them in the same competition is the product of the probabilities of beating them individually. Thus the probability of winning is said to be $P(A) \times P(B) \times P(C) \times \ldots$.

The fallacy lies in the use of the product rule for combining probabilities. The rule assumes that the events to which the probabilities relate are independent. This is not the case because if one competitor has been beaten the probability of beating another is increased. In fact the probability of winning is underestimated by the method. For example, if the probability that A will beat two competitors B and C individually is $\frac{1}{2}$, the probability that A will beat both together is not $\frac{1}{2} \times \frac{1}{2}$ but $\frac{1}{3}$. This is because there are 6 equally likely outcomes of the competition, in 2 of which A wins. They are as follows:

```
A   A   B   B   C   C
B   C   A   C   A   B
C   B   C   A   B   A
```

Thus the probability that the outcome will be one in which A wins is $2/6 = 1/3$.

7.5 The *D* curve method

Although strategies based on data relating to individual competitors are usually unworkable, it should be practicable to improve bidding performance by studying one's own results in relation to the winning bid, assumed here to be the lowest bid. There is a slight theoretical advantage in using the median bid instead of the lowest and making a final adjustment to estimate the lowest, but it is not worthwhile in practice compared with the directness of a strategy based on the lowest bid.

The rest of this chapter describes a bidding strategy which is a statistical representation of what contractors seem to be doing and which can be used by both contractor's and client's estimators. The strategy is a vehicle for disciplined judgement. It is not meant to replace judgement but merely to put it in its proper place – a supplement to, not a substitute for, calculation.

7.5.1 The D curve

The strategy is anchored to a cost estimate which, in principle, excludes any allowance for market conditions. This cost estimate must be calculated by a consistent method. Whether or not it is actually measuring cost is less important than that it should be correlated with it as closely as possible. For a main contractor, sub-contractors' quotations count as costs. The average error of the cost estimate does not have to be zero; neither does it have to be known. Many will be based on approximate quantities and a price book which may or may not be adjusted for inflation, location and any other factors which can be allowed for. In any case it must not be adjusted for subjectively assessed market factors. Whatever method is chosen it should not be changed unnecessarily. If it is, the basis of the strategy will have to be recalculated retrospectively.

The *D* curve is prepared from the results of a large number of bidding competitions. For each competition the lowest bid is compared with the estimated cost and the percentage difference is called *D*:

$$D = \frac{\text{lowest bid} - \text{estimated cost}}{\text{estimated cost}} \%$$

If the strategy is being operated by a contractor he must exclude his own bids.

This variable *D* is recorded for all contracts and its cumulative distribution obtained. Table 7.1 is an example. This distribution is plotted, as in Fig. 7.1, to produce the *D* curve. *P* is the proportion of competitions with greater than the stated values of *D*. In use it has other meanings as well.

It is possible to buy 'normal probability paper' on which the vertical scale is so graduated that, if the distribution whose cumulative frequencies are being plotted is normal, the points will lie on a straight line. Smoothing is easier using such paper and also it may be found that a straight line is a close fit to the

Table 7.1

$D = \dfrac{\text{lowest bid-estimated cost}}{\text{estimated cost}}$ %	Number of competitions	Cumulative frequency	Cumulative % P
−40%	1	76	100
−30%	4	75	99
−20%	6	71	93
−10%	9	65	86
0%	23	56	74
10%	14	33	43
20%	7	19	25
30%	6	12	16
40%	3	6	8
50%	3	3	4
60% and over	0	0	0
	76		

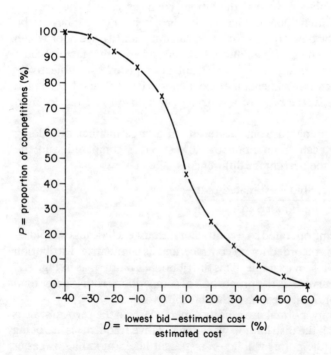

Fig. 7.1 A cumulative frequency curve (the D curve).

points. If that is so, it indicates that the distribution is normal so the percentage points can be obtained from the table of the normal curve once the mean and standard deviation of the original data have been calculated. The normal approximation would be especially attractive if a computer or certain types of programmable calculator were to be used. They could calculate the percentage points from the mean and standard deviation.

In principle the strategy consists of deciding on an appropriate value of P to represent keenness and reading off the corresponding value of D. This percentage is added to the cost estimate to give the bid or estimate.

For a contractor, P represents his desired probability of winning the competition, bearing in mind that the higher the value he chooses the lower the profit he makes if he wins. This trade-off can be made the subject of a separate simple strategy. Otherwise he can repeat the process of using the bidding strategy with different values of P and calculate for each his expected profit. Expected profit is calculated by multiplying the probability of winning by the profit if the bid wins. The probability of winning must be expressed as a number between 0 and 1 by dividing the percentage by 100.

For a client's estimator, P represents his desire to avoid errors in one direction more than the other. If he has no preference he will use $P = 50\%$. If underestimates are twice as objectionable as overestimates he will use $P = 33\frac{1}{3}$. In fact these values of P are not used as they stand. First they have to be adjusted for the tendering micro-climate. The macro-climate, which applies to all competitions at that time, is allowed for after the D curve has been used.

7.5.2 Micro-climate

Small adjustments to P are made to reflect what is believed to be the level of keenness of the bidders for the particular contract in comparison with keenness in the general run of contracts at that time. It will depend on the attractiveness of the contract. If it is thought that bidding will be especially keen P should be increased. If it is thought that it will be less keen than usual P should be decreased. The adjustment should be small and will seldom need to exceed 5%.

7.5.3 Macro-climate

The general or background level of market keenness is measured by the current average value of D, the variable upon which the strategy is based.

$$D = \frac{\text{lowest bid} - \text{estimated cost}}{\text{estimated cost}} \%$$

The value of this variable for every competition should be plotted against the date of the competition to provide an index. Movements of this index would be far too erratic to be clearly indicative, so instead of plotting the individual values they must first be smoothed. Exponential smoothing is the appropriate method and has already been described in Section 3.4.3. Its purpose is to obtain a new index as a weighted mean of the old index and the new value of D.

The choice of smoothing weights depends on the rate at which data arrive. Experiments with the base data trying various weights will quickly show which smooth the movements enough without suppressing the response. The larger weight should be on the old index, so 0.9 and 0.1 would be reasonable ones to try first. It may be necessary to put even more weight on the old index; perhaps as much as 0.95 in a field where results are unusually erratic. If a sudden change in the market is expected, the balance of the weights should be temporarily shifted for a few calculations to quicken the response to the changed level. As soon as the new level is reflected the weights must revert to their usual values.

No 'age correction' is necessary because there should be no long-term trend in the D index, but there is no objection to using one if it is decided that a medium-term trend should be reflected fairly promptly. The method was described in Section 3.4.3.

To adjust the value of D for macro-climate after reading it from the cumulative frequency curve it has to be increased by the amount by which the current value of the smoothed D index exceeds the median (i.e. 50%) value of the curve. In Fig. 7.1 the median value is $+8\%$ so if the D index is running at $+10.1\%$ a value of D read off as -5% will be adjusted to -2.9%.

7.5.4 Number of tenders

Both in the production and use of the D curve the bid should be adjusted for the number of tenders. A greater number of tenders tends to reduce the lowest bid. To calculate the amount of the reduction, use can be made of the shape of the distribution of bids for a contract. Apart from a negligible degree of upward skewness it is closely represented by the normal distribution. This means that if we know its standard deviation we can calculate the effect of the number of tenders as half of the effect on the range. The range, expressed as a proportion of the standard deviation, is given in Table B.

To measure the effect on the lowest bid of changing the number of tenders, the ranges corresponding to each number of tenders can be read from Table B and halved. The difference between the half-ranges for the two numbers of tenders gives the average difference between the lowest bids. Range is expressed in Table B as a proportion of the estimated standard deviation of the population of bids. This is obtained by applying the estimated population coefficient of variation, which is a percentage, to the median for the particular competition. The median is chosen because there are occasional erratic bids and because the distribution is slightly skewed, but the skewness is so little that the arithmetic mean could be used if more convenient. This could be so if a computer is employed.

The population coefficient of variation can be estimated from all available competitions by expressing each bid as a percentage of the mean, or median, for that competition and calculating their overall coefficient of variation. This method has the advantage that the shape of the distribution can be examined at the same time. Otherwise calculating the individual coefficients of variation and averaging them is just as satisfactory.

By this method, before calculating the D values for the purpose of drawing the D curve, the lowest bids can be adjusted to their equivalents for a standard number of bids. The standard can be any convenient number, but it saves effort if it is the commonest number encountered.

The method of adjusting D for the number of tenders should also be applied, in reverse, when the strategy is in use. It should be applied when a value of D has been read from the cumulative frequency curve and has been adjusted for macro-climate. The value of D relates to the standard number of tenders and can be adjusted for the number in the current competition by the following procedure. Unfortunately the explanation is necessarily elaborate and some may prefer to skip straight to the formula which follows it.

The difficulty is that, because the competition has not yet taken place, the median bid is not known. It is required to convert the estimated population coefficient of variation to the standard deviation when only the estimated cost and the value of D for the standard number of tenders are known. These must be used to estimate first the lowest of the other competitors' bids and from that the median bid. The best estimate of the expected value of the lowest bid is obtained from the D curve by reading off the value of D corresponding to a P value of 50% adjusted for the micro-climate. This value of D is used to adjust the estimated cost to give the expected value of the lowest bid for the standard number of tenders.

To estimate the median bid from the lowest bid, apply the current estimated population coefficient of variation to half the range for the standard number of bids in Table B, subtract this product from 1 and divide the result into the expected value of the lowest bid.

The estimated population coefficient of variation can be applied to this estimate of the median bid to give the estimated population standard deviation. With this the effect of changing the number of tenders can be found from Table B by multiplying it by half the difference between the ranges for the two numbers of tenders.

A symbolic representation of the calculation is as follows:

$V\%$ = estimated population coefficient of variation of bids for the same contract.

E = estimated cost using a consistent method.

$D(50+m)\%$ = the value of D from the D curve corresponding to a value of P of 50% adjusted for the assessed micro-climate.

120   *Statistical Methods for Building Price Data*

$R(S)$ = the value read from Table B for the standard number of tenders.

$R(N)$ = the value read from Table B for the actual number of tenders.

The addition to the lowest bid for changing the number of bids from S to N is:

$$\frac{\{[R(S) - R(N)]/2\}\, E\,[(1 + D(50 + m)/100)]\, (V/100)}{1 - [R(S)/2]\, (V/100)}$$

This formula has to be used frequently so it would be worth writing a program for a calculator or a computer.

Some people believe that increasing the number of tenders called for increases the number of cover prices, described in Section 7.2. They will want to reduce the value of N used in the above formula to the number of bids which they believe are genuine. It is the author's opinion that such effects are less than is commonly believed and that N should not be reduced by more than 1 or 2.

7.5.5   Updating the population coefficient of variation

The population coefficient of variation changes much less than D and can be assumed constant if little refinement of the strategy is required. But if effort can be spared it ought to be monitored. The best way is to keep a record of the estimated population coefficient of variation for each competition. Exponential smoothing with heavy loading of old values will provide a figure for adjustments for number of bids and for use in the calculation of D if the assumption of normality has been made. It will also indicate when it is necessary to redraw the cumulative frequency curve.

The best smoothing weights would probably be the same as those used for monitoring D. Although the variability of the coefficient of variation is less, heavily back-loaded smoothing is indicated because quick response to change is not required.

7.6   Summary of the *D* curve method

Estimate the population coefficient of variation of bids using all available competitions, at least 30. The easiest way is to estimate it from each competition individually and average the estimates. From each competition exclude own bids and use Table B to adjust the lowest bid for the number of tenders to give the equivalent for a 4-bid competition, if 4 has been chosen to be the standard number. This requires that the standard deviation be obtained by applying the estimated population coefficient of variation to the median, or arithmetic mean, of the bids in the competition.

For each competition calculate

$$D = \frac{\text{lowest bid} - \text{estimated cost}}{\text{estimated cost}} \%$$

excluding own bids.

Form the cumulative frequency distribution of the adjusted values of D. Plot it on normal probability paper, if available. Smooth the cumulative frequency curve to obtain the D curve. Its horizontal axis is D and the vertical axis is renamed to become the desired probability of winning (P).

Decide whether to use the D curve as it stands or instead to approximate it by a normal distribution. If the normal approximation is to be used, calculate the arithmetic mean and the estimated population standard deviation of D. If preferred the median can be used instead of the mean and instead of calculating a standard deviation the percentage points of the frequency curve can be converted to a standard deviation with Table A.

Record the values of D in date order and exponentially smooth them to give a D index which is kept up to date. Similarly, exponentially smooth the estimated population coefficients of variation and keep this average up to date.

7.7 Using the strategy

Calculate an estimated cost using the same objective method as usual when calculating D. If a contractor, decide on as low a probability of winning as is acceptable. Two or three probabilities may be chosen and parallel calculations made. The one finally selected would be the one which maximized profit when the products of profit times probability were compared. If a client's estimator, choose a value of P which represents the desired probability of underestimating. Adjust the desired probability by an amount representing the assessment of the micro-climate. Look up the value of D corresponding to the chosen, adjusted probability of winning (P).

If the normal approximation is being used, use Table A backwards by regarding the adjusted probability of winning as the area under the normal curve in the body of Table A. The corresponding margin figure is the number of standard deviations the required value of D is from the mean or median. Therefore, multiply it by the estimated population standard deviation of D calculated in the preparation. Subtract this product from the median of the D distribution to give the value of D.

If the figure obtained from Table A is negative, as it will be for probabilities of winning of less than 50%, subtracting the product will be adding to the median.

Whichever method of obtaining D was used, adjust D for the macro-climate by adding to it the amount by which the current smoothed D index exceeds the value of D when P is 50%.

122   *Statistical Methods for Building Price Data*

Calculate the bid suitable for the standard number of bidders from this adjusted value of D as follows.

$$\text{Bid} = \text{estimated cost} \left(1 + \frac{D}{100}\right)$$

For the purpose of calculating the bid the estimated cost must be the one calculated objectively by the standard method. A more refined or appropriate estimate of cost could be used for other purposes, such as calculating profit.

Finally, adjust the bid for the number of bidders in the competition using the current exponentially smoothed average estimate of the population coefficient of variation in the formula in Section 7.5.4. It is important to do this adjustment for the number of tenders last, so as to minimize the recalculation necessary if the number has to be changed.

7.8 Example

7.8.1 Preparation

The strategy will be based on the 76 competitions for a particular type of work which are the subject of Table 7.1. For each competition the population coefficient of variation of the bids in the competition has been estimated by using the $\sigma(n-1)$ button on a calculator, dividing by $\bar{x}$ (the arithmetic mean calculated at the same time) and multiplying by 100. The arithmetic mean of these 76 estimates of the population coefficient of variation was 6.8% and their median was 8.2%.

The most recent 20 figures were (%):

11.4 5.9 5.8 6.9 7.3 5.4 6.7 6.8 7.9 10.2
8.1 6.0 4.9 13.8 10.1 3.9 5.8 6.5 8.3 7.0

Own bids were included.
For each competition the value of

$$D = \frac{\text{lowest bid} - \text{estimated cost}}{\text{estimated cost}} \%$$

has been calculated using a consistent estimating method. Own bids were excluded. Adjustments of lowest bids to a standard number of tenders of 4 had already been made.

The D curve has been drawn (Fig. 7.1). Although the frequency distribution is reasonably symmetrical the frequencies are more concentrated at the centre than in a normal distribution. This rules out the use of the normal approximation so the D curve will be used.

The reason for this shape may be a change in the dispersion of the D values so that the distribution is a mixture of distributions. However, this is only a guess and until it becomes clear that the D curve should be based on only some of the data it is best to use all of them as they stand.

Tender patterns and bidding strategy

The D values for the latest 20 competitions were (%):

-8 6 4 13 3 0 -2 3 40 8
2 -10 31 5 12 10 1 22 6 18

Exponential smoothing of the values of D need only be done on the latest, say 20, values provided that a very high or low starting 'old' value is not used. In this case an old value of 4% will be assumed; it being the average of the first 5 values in the above 20. Weights of 0.9 and 0.1 will be used. The smoothed values are in Table 7.2.

Table 7.2

$4 \times 0.9 + (-8) \times 0.1$	= 2.8	
$2.8 \times 0.9 + 6 \times 0.1$	= 3.1	(in fact 3.12 is held in the calculator and used in the next line)
$3.1 \times 0.9 + 4 \times 0.1$	= 3.2	
$3.2 \times 0.9 + 13 \times 0.1$	= 4.2	
$4.2 \times 0.9 + 3 \times 0.1$	= 4.1	
$4.1 \times 0.9 + 0 \times 0.1$	= 3.7	
$3.7 \times 0.9 + (-2) \times 0.1$	= 3.1	
$3.1 \times 0.9 + 3 \times 0.1$	= 3.1	
$3.1 \times 0.9 + 40 \times 0.1$	= 6.8	
$6.8 \times 0.9 + 8 \times 0.1$	= 6.9	
$6.9 \times 0.9 + 2 \times 0.1$	= 6.4	
$6.4 \times 0.9 + (-10) \times 0.1$	= 4.8	
$4.8 \times 0.9 + 31 \times 0.1$	= 7.4	
$7.4 \times 0.9 + 5 \times 0.1$	= 7.2	
$7.2 \times 0.9 + 12 \times 0.1$	= 7.6	
$7.6 \times 0.9 + 10 \times 0.1$	= 7.9	
$7.9 \times 0.9 + 1 \times 0.1$	= 7.2	
$7.2 \times 0.9 + 22 \times 0.1$	= 8.7	
$8.7 \times 0.9 + 6 \times 0.1$	= 8.4	
$8.4 \times 0.9 + 18 \times 0.1$	= 9.4	

If it is decided that weights of 0.95 and 0.05 should be tried, it would be necessary to cover the latest 40 values. (A reasonable number is twice the ratio of old weight to new, i.e. $2 \times 0.95/0.05$.) It seems that the D values had been in a low period but may have returned to values close to the median of all 76 which was 8.2%.

Exponential smoothing of the latest 20 estimates of the population coefficient of variation would begin as follows. The first old value is, again, the average of the first 5 values:

$$7.5 \times 0.9 + 11.4 \times 0.1 = 7.9$$
$$7.9 \times 0.9 + 5.9 \times 0.1 = 7.7$$
etc.

and ending with 7.4.

7.8.2 Use

The competition for which a bid is required is one which the contractor would like to stand slightly more than the usual chance of winning. This is because he has a comfortably full order book but could just squeeze this job in if he slowed down some others. The work is of a type that he likes, the client is a reasonable one and the architect easy to work with. These influences almost cancel each other out but leave a slight balance in favour of more than average keenness.

The contractor usually hopes to win about a fifth of the competitions for which he enters, so for average keenness he would use a desired probability of winning of 20%. In this case he decides to use 30%. He guesses that the contract would be fairly popular with other bidders so he needs to enter the D curve with a P value slightly higher than 30%. He decides that the micro-climate requires it to be raised to 33%.

The value of D corresponding to a P value of 33% is 14%. The current value of his D index (the exponentially smoothed average of the values of D which he calculates from the results of all competitions for which he estimates) was calculated above and stands at +9.4%. This is 1.2% above the 50% point on his D curve so this must be added to the 14% so far calculated.

The total of 15.2% is the amount by which his bid should exceed his estimated cost, so his bid would be $E + 0.152E = 1.152E$ if the number of other tenders were 4. He believes there are 6 other contractors intending to bid. To preserve his desired probability of winning he will have to reduce his bid. The amount is calculated as follows. He may wish to repeat it for other possible numbers of tenders.

The latest exponentially smoothed average of the estimates of the coefficients of variation of the population of bids is 7.4%. To use the formula given at the end of Section 7.5.4, the following are required:

$V = 7.4\%$

$D(50+m)\% = D(53)\%$ including the allowance for micro-climate)

$\qquad = 5\%$ (reading the value of D corresponding to $P = 53\%$)

$\left. \begin{array}{l} R(S) = R(4) = 2.06 \\ R(N) = R(6) = 2.54 \end{array} \right\}$ from Table B

The amount by which he should change his bid to allow for there being 6 other bidders instead of the standard 4 is:

$$\frac{\{[R(S) - R(N)]/2\} E [(1 + D(50+m)/100])\ (V/100)}{1 - [R(S)/2]\ (V/100)}$$

$$= \frac{-0.24 \times E \times 1.05 \times 0.074}{1 - 1.03 \times 0.074} = -0.020 \times E$$

Therefore his bid should be reduced from $1.152E$ to $1.132E$.

Further calculations of his expected profit and other financial considerations may make him less keen on winning than he originally was. In that case he should rework the calculation with a lower level of keenness, say 25% instead of 30%. This would produce a bid of $1.172E$. This may give him sufficient profit to justify a level of keenness of 25%. If not, perhaps a further reduction to 20%, producing a bid of $1.212E$, would be acceptable.

Experience of using the D curve would allow an acceptable harmony between profit and keenness to be achieved quickly.

It may often be the case that a contractor would be less keen on winning a contract if he realized how much profit he could make with an acceptably lower probability of winning. He may be willing to enter a larger number of competitions and rely on chance to give him the necessary proportion of wins.

7.9 Further development

There is more that could be done to extend the strategy to maximize profit and to take into account costs incurred by estimating. The method described here could form the core of a comprehensive strategy suited to the firm's policy.

A computer program could be written which required as input only the amounts tendered by all bidders in previous competitions, the estimated costs and smoothing weight changes to provide the basis of the strategy. In use the inputs would be the desired probability of winning, the assessment of the micro-climate and the number of other bidders.

The program could be extended to cover other factors, such as financial considerations, but it is important that it should output intermediate stages. It must not become a black box whose working is not understood by the user, because he must be able to see when an assumption upon which the method is based no longer holds.

7.10 Testing and tuning

Any new strategy should be tested on past data and then run in parallel with existing methods. It is not enough to judge the performance of a strategy using the same data as that upon which it was based. Tests on such data would show the strategy in too good a light.

The first requirement is to see whether its performance is as good as existing methods. If not, it must be considered whether this is because its performance is more erratic or whether its average is wrong and it merely requires a simple addition or subtraction from all its results. The appropriate analogy is with the rifle. If it is sufficiently consistent it can easily be zeroed. Tuning, in the case of this strategy, consists of getting used to choosing values of P which truly reflect the bidder's wishes.

When assessing the success of this or any other bidding strategy (or any estimating strategy) it is important not to pay attention to individual results. There are bound to be chance occasions for a contractor's estimator when the tender was exactly right (just below the second lowest) or, for a client's estimator, when he was very close to the lowest tender. Although these may be good moments to ask for a pay increase, they do not alone indicate much about the value of the strategy. Only long-run results based on carefully kept records can form a basis for accepting, rejecting or retuning a strategy.

The best method of assessment is to run the two methods, new and existing, side by side for at least 40 competitions. The criteria for the comparison must be decided in advance and, for contractors' estimators, it is not obvious what they should be. The best is probably total profitability, but contractors have other objectives as well and each would have to decide how best to measure the extent of his achievement.

8 Cost models

8.1 Types of cost model

It is tempting to pay too much attention to defining cost models and generalizing about them. In fact, little need be said about their principles. Any way of estimating cost, or estimating the effect on cost of changing a building's characteristics, can be described as a cost model. The term is usually reserved for unconventional methods which are less direct than applying cost rates to measurements. But if the bill of quantities were not already in common use it would certainly be described as a cost model.

A model is intended to respond to an input by producing the same output as the real thing. The input to a cost model is information about a design and perhaps some cost constants, although these may be embedded in the model. The output is cost. Discussion of cost models is discussion of ways of representing costs. This is a big subject so it will be dealt with here only in those respects which have a statistical connection. Attention will be focused on capital cost but the principles apply equally to cost in use.

There are two distinct ways of representing costs: the realistic and the 'black box'. Realistic methods are derived from attempts to represent costs in the ways in which they arise. 'Black box' methods do not attempt to represent the ways in which costs arise. Their only justification is that they work. As they are further developed the realistic methods lose some of their contact with reality because of difficulty in obtaining data. As black box methods are developed they are rationalized by explanations of some of their relationships in realistic terms. Development brings the two types closer together but the distinction remains important to an understanding of their underlying strengths and weaknesses.

The following treatment moves from realistic to black box methods, passing through those commonly used and which are called here 'in-place materials related' and 'area-related' methods.

8.2 Realistic methods

The most realistic method identifies the direct causes of cost and measures them directly. If a contractor plans the whole construction process in detail, he can estimate his cost by totalling the materials, plant, labour and other costs. All of these relate to specifically described items, such as actual plant and gangs of men, employed for measured lengths of time.

Such a method has the best potential accuracy but, even for a contractor, may be too laborious for all tenders. The client's cost adviser is further inhibited by lack of detailed information about methods of construction and about the costs of the items.

Although difficult with existing facilities, the estimator should keep in mind the desirability of staying as close as possible to this realistic representation of costs. The increasing availability of computers makes it possible to consider detailed methods which previously had been rejected as too cumbersome. Even problems of obtaining data can be eased when a method of using them has been developed, because it becomes more worthwhile to provide a central data bank. Also, the data bank can be made self-generating if each use adds data for another project.

The methods for computer simulation of industrial production operations are already known. It is likely that eventually computer programs will be developed far enough to apply to the production of buildings. Simulating construction before beginning on site will encourage the standardization of construction methods which has for so long been predicted but which is so slow in arriving. When actual construction is routinely simulated the calculation of costs will be a fairly simple by-product.

Meanwhile, even if the process of construction is too complex, the simulation of pre-contract planning methods is possible using the techniques of expert systems in which the decision processes of an experienced practitioner are simulated by a computer program. In fact this type of simulation is most likely to predict what contractors believe their costs to be until simulation of actual construction becomes normal practice in the industry.

Such a computer program was devised and initially developed by the author in 1971 at, what is now the Property Services Agency of the Department of the Environment. It has been further developed at the University of Strathclyde. It is called COCO (costs of contractors' operations). It is based on decision criteria governing choice of construction methods and choice of plant, performance data for plant and labour and procedures for harmonizing work rates of plant and labour and for smoothing resources. It was found that the planning methods of large contractors differed from each other only a little. Where there were differences it was possible to simulate them separately and produce a range of results.

A simulation of contractors' planning decisions requires a few qualitative data about the design such as bay size, to determine what sort of formwork can be used, and the position and weight of heavy indivisible items to be lifted by a crane. A little information about the site, including restrictions on positioning cranes and the cart-away distance for spoil, is also required. To calculate plant and labour requirements needs some quantity data of course, but how much depends on how accurate the resource calculations need to be. Most of the resources are determined by the quantities of a few key materials and a few dimensions. For calculation of materials costs more data are required.

One reason for preferring a realistic method is that it can show the designer the construction consequences of his design decisions, and easily constructed buildings attract keener prices. For instance, it may be known quite early in the design process that there will have to be very big precast concrete cross beams and that an unusually large crane will be needed to lift them into place. The cost effect of the large beam, and thus of the design feature which makes it necessary, can be readily calculated. Realistic costing may show that some of the cost may be offset by the rest of the design being affected by the presence of a crane capable of lifting heavy items. Perhaps larger cladding panels can be specified without extra cost.

The use of a realistic costing method (COCO) before the design has been finalized has brought to the notice of designers potential construction problems. Dealing with them later, after a contractor had pointed them out, would have been difficult and not dealing with them at all would have produced higher bids from the contractors.

Moving from simulation of the construction process to representing the planning process is one departure from realism; though not for the client's estimator who is trying to estimate what the contractor believes that his costs will be. Another departure may be necessary if the work of sub-contractors cannot be simulated. This could be so if they are specialists who are unwilling to reveal their methods to researchers or whose methods vary too much for generalization. It may be necessary to deal with them by the more usual in-place materials related methods.

8.2.1 Data banks for realistic methods

For the contractor there is no difficulty in obtaining the necessary data for realistic costing of his own operations. As he calls for quotations he updates his cost data for materials and plant hire. Even if the plant is his own, he has notional hire charges. Labour costs have to be recorded in any case. The client's cost adviser has to rely on published sources and special research. If the costing method is established there will have to be a body responsible for maintaining the data bank. In relation to sub-contractors the main contractor is in the same position as the client's cost adviser. He must rely on published costs and the work of those who maintain the data bank.

Another attractive method is to store data about projects in the form of resourced networks or linked bar charts. Such data could be collected from subscribing contractors by the staff of the data bank and put into standard forms for storage. Their most difficult problem would be to relate the networks to features of designs so that appropriate networks could be retrieved from the bank and the dimensions varied.

Data could be used by subscribing contractors to help them synthesize cheaply a plan for a project before tendering. They could calculate a standard cost for the job which they could adjust to provide a bid. Bidding strategy would be more reliable and keener bids would be encouraged. The plan and data would also be a useful starting point for planning construction if the contractor obtained the job.

The data bank could also be accessed by subscribing cost advisers and thus ensure that the advice they gave tended to persuade designers towards economical methods. This would be welcome to contractors. Most of them prefer to have easily built designs.

Eventually the common use of such a network bank could lead to an agreed basis for costing which could replace the bill of quantities or confine it to a schedule of materials. Variations and claims could be calculated and settled on the basis of the agreed construction plan.

8.2.2 Variability in realistic methods

Variability between construction or planning methods has already been mentioned. The range of results produced if all are simulated does not seem to be large but has not been accurately measured. It would be instructive to run such a simulation for several contractors on each of many projects using the same resource costs so as to measure this cause of variability between contractors' costs.

The other cause is variability between contractors in their resource costs. This could be measured by comparison of contractors' records and built into the data bank as coefficients of variation attached to the mean costs. They are likely to be normally distributed.

Combining the variabilities of resource costs presents a problem. In such cases a Monte Carlo method is often used. Whenever a resource cost is required for an item a value is drawn at random from the distribution of costs for the item. This is repeated for all items and a total cost obtained. This is repeated a great many times to give a distribution whose variability can be measured. A possible objection is that this method of measuring variability requires the assumption of independence between items of resource cost. If a contractor with high costs for one item tends to have high costs for other items, this increases the variability of total cost and the Monte Carlo method underestimates it. However, if this is thought not to be so, an assumption of independence may be made. Then the Monte Carlo method is the best way of dealing with the variability, and a computer makes light work of it.

The Monte Carlo method for a normal distribution is applied by using the mean cost for each item plus what is known as a random normal deviate. This may be a facility of the computer's compiler so that something like 'GAUSS

(8.2, 2.1)' can be an instruction in the program. This would mean 'provide a randomly chosen value from a normal distribution which has a mean of 8.2 and a standard deviation of 2.1'. The mean and standard deviation relate to the item being considered. If the compiler has no such facility it will almost certainly have a random number generator so that the following formula can be used.

Random normal deviate with mean $\bar{x}$ and standard deviation σ

$$= \sigma \sqrt{(-2 \log_e R)} \times \cos(2\pi \times S) + \bar{x}$$

where R and S are separately obtained random numbers in the range 0 to 1. R and S can be, for instance, two-digit random numbers divided by 100 or three-digit random numbers divided by 1000.

When the costing process has been gone through for all the items for which random normal deviates were required, the total cost is noted and the process repeated. A different cost will result and it is the distribution of the total costs which provides the basis for setting confidence limits on the total cost. The values within which 95% of the totals of the random costs lie are the 95% confidence limits.

The best estimate of the total resource cost is the total of the mean costs for the items. As a check on the application of the Monte Carlo procedure the mean of the totals should be compared with the total of the means.

If the assumption of independence is thought to be unwise the Monte Carlo method must be modified to reflect the tendency for high cost for one item to be associated with high cost for another. This can be done by producing the randomly chosen cost in two stages. First a single random normal deviate, with a fairly small standard deviation, is obtained and this is added as a percentage to all the means. Next the Monte Carlo process is gone through as before but with a slightly reduced standard deviation. The square of this reduced standard deviation and that of the common addition must add up to the square of the standard deviation obtained from the data bank for each item.

The size of the standard deviation of the common addition is a matter of judgement. It is unlikely that data could be obtained to calculate it. A practical way out of the difficulty is to go through several test cases in which the correct result is known. Varying the size of the common addition will soon tune it sufficiently accurately.

8.3 In-place materials methods

Methods using unit prices for described items of work, in-place materials methods, are a modification of realistic methods and are the commonest detailed costing methods in use. There is no need to enlarge on them here except to remark that if there were no computers and cost manipulators had only extremely laborious realistic methods available, the idea of saving work by attaching plant, labour and other costs to the quantities of materials would seem brilliant.

Unit prices provide a common currency and are well understood. They must not be abandoned precipitately but consideration should be given to methods which were too laborious or complex before computers were commonly available. An essential difference between realistic models and those which attach costs to quantities of work is that the former allow resources to be smoothed whereas the latter assume that the amount of plant and labour can be varied to match exactly the progress of the work. Admittedly an average allowance for the cost of smoothing is implied in the unit prices but this is not varied from one job to another. For estimating purposes this difference may be unimportant for fairly simple construction, but for the designer to be advised on how to simplify, and therefore cheapen, construction realistic models are required.

8.3.1 Variability in in-place materials methods

The variability of unit prices for work items can be measured directly from a sufficient number of projects and has already been dealt with in previous chapters. Monte Carlo methods are sometimes suggested for evaluating the total variability for a combination of items with differing variabilities. Whereas it may be reasonable to assume independence between items of resource cost, it is certainly not so for costs of items of work in in-place materials related methods, where the same resource occurs in many items. This produces correlations between projects which completely invalidate the assumption of independence and increase the variability of estimates based on average unit rates. The method of simulating dependence described at the end of Section 8.2.2 could be used but the arbitrary common standard deviation would have to be large and the result would be very sensitive to the chosen value.

An estimating method which averages separate estimates, each based on one price analogue, does not suffer from this disadvantage. In fact, the negative correlations between unit prices actually reduce variability.

8.3.2 Data banks for in-place materials models

For in-place materials related models the basic data are unit prices for items of work. These are gathered from actual projects so there is no difficulty in measuring variability. Whereas in data banks for realistic models the data come from various sources and not necessarily from identified projects, in the case of unit prices they can be kept in sets so that each set relates to a single project.

Cost models

To measure the variability of the estimates from such a data bank there is no need to use a modified Monte Carlo method. Each set of unit prices can be multiplied by the appropriate quantities to produce an estimate from each set. The amount by which the estimate of price could vary if it followed past variability is then given directly. Prices for minor items which do not appear in a set can be substituted from another source. If very large items are missing the set cannot be used in this way, unless great reliance can be placed upon judgement. It is unlikely that the estimate of variability will be seriously affected.

The following example shows how the variability in the estimates can be measured. Suppose that from a data bank of unit prices, data relating to projects fitting a broad description have been extracted. The selection has been made without knowing what the prices were. The description relates to a project for which an estimate is required and for which quantities are available. The description has not been made so specific that too few cases are qualified. It merely specifies the type of building in broad terms and a minimum floor area to eliminate the smallest.

Eight projects answered the description. The unit prices for each project have been applied to the quantities in the project for which the estimate is required. This produces an estimate from each of the eight. Predetermined adjustments have been made to allow for different dates, locations, percentages of preliminaries, prime cost sums and any others previously decided upon. The resulting estimates in ascending order are:

200 881
209 315
212 246
226 902
227 817
239 549
248 387
252 411

To measure the variability of the estimates there is a choice of methods. Quantiles (in this case quartiles would be suitable) can be obtained directly and may be sufficient to convey the variability. Alternatively, the standard deviation can be calculated and any desired percentage points obtained from Table A on the assumption that the distribution is normal.

A measurement of the variability of the estimates is likely to be required for its own sake to compare with previous experience. Also, quantiles can be used to choose an appropriate level at which to pitch the final estimate to give the desired probability of being too high or too low. If the probability of being high is required to be twice the probability of being low it would be reasonable to choose as the estimate a figure which is $3\frac{1}{6}$ of the way through the eight numbers. The number $3\frac{1}{6}$ comes from the formula

$$\frac{n}{q} + \frac{1}{2}$$

where n is 8 and q is 3 because a third of the distribution is to be greater than the figure. This quantile is the $66\frac{2}{3}$ percentile. In the above example this is 237 594, which would be rounded to 238 000.

A different requirement might be to estimate how far the percentile could differ by chance from the same percentile of the parent population of such estimates. For this the standard error of the percentile is required. First the standard deviation of the eight estimates (regarded as a random sample of the parent population) is calculated, incorporating Bessel's correction by using the $(n - 1)$ button on the calculator. This is divided by $\sqrt{8}$ and multiplied by a factor depending on which percentile is being considered. For a 50% or lower percentile the factor is $1.12 + 6/P$, where P is the percentile. For percentiles above 50% the factor is $1.12 + 6/(100 - P)$. In the example P is $33\frac{1}{3}$, so the factor is 1.30.

The estimate of the standard deviation of the population is 18 798. Dividing this by $\sqrt{8}$ and multiplying by 1.30 gives 8640 as the standard error of the percentile. This gives an idea of how much the limited sample size might have affected the chosen figure. It could possibly be that if a very large sample could be taken, the $66\frac{2}{3}\%$ point would be as high as 238 000 + 14 000 (1.64 standard errors if we are only interested in one direction). This leads to an estimate of 252 000.

The increase from 238 000 to 252 000 would only be made if in this particular case extra caution were required. A long-run proportion of $\frac{2}{3}$ high estimates can only be obtained if the percentile is used as first calculated.

8.4 Area-related methods and cost planning

Area-related methods are another step away from realism. They not only make the assumption that work is related to materials in the same way in different designs, but also that they are both related to area. In its crudest version the relationship is assumed to hold throughout a given type of building.

Cost per unit floor area, unless a functional cost (e.g. cost per hospital bed) is more appropriate, is often used for providing a first estimate for budget purposes. It is also used for setting cost limits for a class of building in order to control quality and efficiency during design.

For cost planning, and sometimes for estimating, the principle of relating cost to area is applied separately to parts of the building (elements). The area is not always floor area but area of the part being estimated. Thus the designer can compare the estimated cost of parts of his design with cost targets to judge where economies are most likely to be found. Although the methods of cost planning have been refined, such detailed cost control often assumes more accuracy in estimating than is possible with present methods. The coefficient of variation of estimating errors in whole building prices can be measured by comparison with accepted tenders. Morrison and Stevens (reference 1 at end of chapter) found that it is seldom less than 10%, even in specialist work. Considering that the variability of percentage errors in estimating parts of the building is likely to be greater, the foundation for cost advice to designers is shaky.

Where the primary purpose of cost control is to control quality, with only a long-run average control of costs over a series of projects, there may be better approaches than simply through costs. Perhaps the type and quantity of the main materials and the construction method and resource implications of the design could be compared with those of previous or standard satisfactory designs to form a judgement of the economy of the design. Then, rather than comparing with a budget, the cost adviser could certify that the design was economical and easily built. The bidding competition would then be relied on to produce an acceptable long-run series of prices. The estimate could become merely one of the guides for the cost adviser and an approximate budgeting figure for the client.

This is a big subject with implications for the way in which standards can be controlled by the client and for the skills required by a cost adviser. A further discussion would be out of place here but, to summarize, in any method of cost planning and control careful thought should be given to how the variability of estimating affects the way in which any proposed means of control can be exercised. If variability is too great for control in individual projects, it may be necessary to accept that long-run average control of costs is the best that can be done and that quality should be controlled separately.

8.5 Area-related methods and estimating

Regarding area-related methods as cost-estimating models, the simplest are developments of the basic relationship between cost and area. Area has such a powerful effect on cost that it needs to be allowed for first before other effects can be discerned. Next, type of building has such a strong effect that it is usual for a separate model to be developed for each type. Type is usually interpreted as function (e.g. schools) but it ought to be possible to classify buildings in a way more relevant to cost. Adjustments for location are likely to be made so this can be regarded as a factor in an area-based cost model. As factors are added they build up a black-box cost model.

Simple models are usually elaborated as time goes by but it is not often that analyses of costs are made to ensure that each extra factor introduced into the model has earned its place by a worthwhile reduction in the coefficient of variation of costs per unit area. This test is necessary if the entry of factors into a model is to be controlled. The effect on variability should be tested by application, with and without the factor, to at least 50 projects. It must be stressed that it is the effect on variability which matters, not the effect on the average. The latter can be adjusted by 'tuning', that is, by adding or subtracting a constant amount from all estimates. This takes the form of a constant in a formula.

This principle of testing before acceptance is more important than sophisticated methods of identifying new factors. There is no harm in subjective selection and quantification of factors for final incorporation in models, providing that testing is objective, cautious and thorough. Also, there is no harm in provisionally inserting in models factors which appeal to reason and which have only been tested to the point of making fairly sure that they do not increase variability.

An example of a model which has been subjectively proposed and objectively tested could be one developed by observation of a relationship between cost per unit floor area and total floor area. Plotting one against the other might produce a clear curved line through the points, as in Fig. 8.1. In the figure the line has been drawn so as to equalize as far as possible the number of points above and below the line for as many sections of the line as possible and for the whole line. This could be called a 'median fit'.

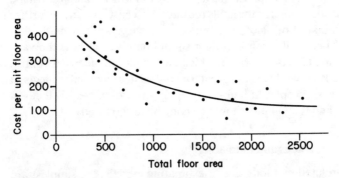

Fig. 8.1 A median fit.

If there is no reason to prefer a formula then the fitted curve can be used directly for adjusting costs per unit floor area to allow for different areas. For computer methods it is much easier to use a formula so there is always interest in expressing the relationship in that way.

In the example in Fig. 8.1 it is clear that a relationship exists, but its form is not obvious. If the curve is drawn smoothly it may be possible to read off some points from it and replot it on logarithmic graph paper. It is just as good to take logs of the area figures and plot them on ordinary paper, as in Fig. 8.2. If either of these methods produces a straight line it can easily be translated into a formula.

In the example illustrated in Fig. 8.2 it can be seen that the line drops from

400 to 100 cost units in 1.0 units of $\log_{10}$ area. For each unit by which $\log_{10}$ area is increased the cost per unit area falls by 300 (within the range of the data).

At $\log_{10}$ area $= 2.4$, cost is 400. For every unit of $\log_{10}$ area greater than 2.4 we must take 300 from cost. The amount by which $\log_{10}$ area exceeds 2.4 is ($\log_{10}$ area $- 2.4$) so we must take 300 times this from 400. This can be expressed as follows:

Cost per unit area $= 400 - 300 \times (\log_{10}$ area $- 2.4)$

Taking the brackets off, this gives

Cost per unit area $= 1120 - 300 \log_{10}$ area

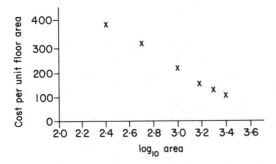

Fig. 8.2 Logarithms of points on the fitted line.

Straight-line relationships are worth seeking but if one does not readily appear then curve-fitting methods can be used. There are curve-fitting methods in most computer libraries. They do not require a line to be fitted by the user and they indicate the type of relationship between the variables which comes nearest to fitting them. The methods are fully described in the manuals. Their disadvantage is that they produce optimum but complex relationships. They are most useful as guides to a simple relationship if one exists.

When several relationships between design factors and cost per unit area have been established they can be used separately as adjustments to previously costed designs or they can be combined into a single model. In either case the danger of correlated factors must be avoided for the following reason.

It is possible that two factors overlap so that adjusting for one is partly adjusting for the other. A relationship between cost per unit area and type of building may be influenced by certain types strongly tending to be of steel frame construction. If adjustment factors for type of frame have also been calculated, applying one adjustment and then the other will overcompensate for any real effect of steel frames on cost.

One way of dealing with this problem if it is detected is to combine the two relationships. If they came from the same data they must be reanalysed in such a way that only one set of adjustment factors is produced for the correlated relationships.

For example, it may have been found that all forms of construction other than steel frame can be treated together, because their costs per unit area were similar, provided that high buildings were separated. The tendency for industrial buildings to have steel frames would then be dealt with by tabulating the data as follows. The numbers in Table 8.1 are not costs but are the numbers of buildings. This is to show the sort of sample sizes required, but in reality median costs per unit area would be entered. Where the number is in parentheses there are too few cases to support averages.

Table 8.1

	Steel frame		Other construction	
	1 or 2 storeys	3 or more storeys	1 or 2 storeys	3 or more storeys
Industrial	21	13	14	9
Commercial	(2)	14	10	25
Domestic	(1)	(3)	19	7

To obtain a factor between any two cells in the table, one cost per unit area is divided by the other. Adjustment factors can be obtained for any number of relationships and data banks can be adjusted to standards. For example, all costs per unit area could be adjusted to what they would be if the building were a single-storey steel framed industrial building. If preferred, the data can be held in the bank unadjusted and all adjustments made at the time of use. The choice of method is entirely a matter of convenience. It is possible to devise an estimating method which is really a series of adjustments to data from the bank done separately or in groups as already described.

Rather than making separate adjustments there is a great attraction in analysing the data bank once and for all to produce a formula covering all factors likely to be of interest. The characteristics of the design for which the estimate is required would be substituted in the formula to give an estimate. The formula would be of a form in which a starting cost per unit area would be provided and then appropriate amounts would be added or subtracted according to the characteristics of the design. Thus the method of use would be the same as when separate adjustments were made. The special feature of the method would be the single comprehensive formula obtained by a single analysis of the data bank. Multiple regression analysis provides this.

8.6 Regression methods

This more organized way of setting up cost models can be used to consider a large number of variables. Multiple regression analysis, or another multivariate technique, can select those which are most likely to be useful. An advantage of multiple regression analysis over some other multivariate analysis techniques is that, at the same time as identifying the variables having most influence, the influence is conveniently quantified. The result is a formula in which each variable has a multiplying factor. For example,

Cost per unit area = 12.95 − 0.002 × floor area

+0.38 × number of storeys + 2.32 if in London

Cost per unit area is called the dependent variable, 12.95 is the constant and −0.002, 0.38 and 2.32 are coefficients. 'Area', 'number of storeys' and 'London' are independent variables, called here 'factors'.

If used thoughtlessly this method is the blackest of black boxes. It is possible to include any variables in the list for testing as factors and, sometimes, one is selected whose apparent effect contradicts common sense. This is not the fault of the technique: it is the responsibility of the user to ensure that a 'straight-line best fit' to the data in the sample is what he wants.

Multiple regression analysis fits a straight line to data in a number of dimensions equal to the number of factors. This is best understood by imagining how a straight line would be drawn through points scattered in not two but three dimensions. They would relate to three factors. It would obviously be possible but very difficult. The mathematical procedures which comprise multiple regression analysis do it for any number of factors but they are too laborious to be carried out for any worthwhile quantity of data without a computer.

The method of fitting the straight line minimizes the 'distances' of the data points from the line. The success in doing this is measured by the residual error in estimating the dependent variable. The dependent variable is usually cost per unit area. In fact the method minimizes the sum of the squares of all the 'errors'.

8.6.1 Outlying data

Already this introduces a way in which multiple regression analysis may be unsuitable for building price data. More importance is given to outlying, and possibly wrong, data than we would wish. This would be true even for 'robust' regression methods which minimize absolute errors rather than their squares. Section 2.9 in Chapter 2 shows that mean deviation is affected by outliers almost as much as is standard deviation.

A more promising development is median regression, which is easy to apply graphically when there is only one independent variable as in Fig. 8.1. The author's method is to choose intervals on the horizontal axis which divide the plotted points into groups of at least 4. In the case of Fig. 8.1 intervals of 500 are suitable up to 2000 but they need not be equal. The medians are marked lightly and a smooth line drawn through them by eye. It is sometimes necessary to widen some intervals before a satisfactory line can be drawn.

There are also mathematical methods in some books on non-parametric or distribution-free statistical methods. The theory for multiple median regression needs to be applied so that it is available in a computer statistical library. Then a powerful tool will have been added to the inadequate equipment so far available to those struggling with cost modelling using data of uneven quality covering several independent variables.

8.6.2 Linearity

Straight-line relationships are assumed, even though they are no more likely than any other. The effect of the assumption can be mitigated by testing for linearity in the relationships of the dependent variable and each of the factors in turn. If the relationship is clearly nonlinear the factor can usually be expressed differently in a way which makes it approximately linear. This process of transformation is described in Sections 4.3 to 4.5. Care must be taken not to 'over-transform'. If the correction seems to be too strong it is better to use a weaker one, such as square root, or none at all.

Something else to be watched for is whether the scatter about the line is roughly the same thoughout its length. The appropriate transformation should remove any strong tendency for the variability to increase or decrease from one end of the line to the other.

Whether or not a transformation is used, linearity cannot be assumed beyond the limits of the data unless, in rare cases, a 'natural law' is believed to be operating. For example, it may be established that a calculable amount can be added to the cost per unit area for each extra lift to be installed. If so, the user may be willing to assume that the relationship holds for larger numbers of lifts than actually existed in any of the buildings for which he has data. Such circumstances are rare. It is usually unwise to assume that relationships hold beyond the limits of the data, and often not even as far as that when the limit is represented by a single case.

A few experiments with altered data should be conducted to familiarize the user of a multiple regression program with the effects of isolated 'points'. Outlying cases can have a strong effect on the fitted line so the analysis should be repeated without such cases to show whether the value of the coefficient for a factor depends heavily on them.

Before starting a multiple regression analysis a great deal of background information is obtained by plotting the dependent variable (cost per unit area) against each factor in turn. This should always be done to aid decisions on linearity, the need for transformation and the influence of outlying cases. The plotting need not be precise and can be quite rapid.

A computer will be necessary for analyses large enough to be useful, so advantage should be taken of its speed, cheapness and ease of repetition to run several analyses with different variables and transformations.

8.6.3 Correlations

The effect of correlations on adjustment factors has been mentioned in Section 8.5. They have the same influence on multiple regression analysis. They can be detected by examining the correlation matrix which is part of every program. It prints the correlation coefficients between every pair of variables. When two independent variables are highly correlated their joint effect will be properly evaluated but the individual coefficients may be very unreliable. The best solution is to combine them in some way. If the two variables are number of storeys and total floor area a suitable combined variable is floor area per storey. If that is used one of the other two can be used as well.

It is wrong to have too many factors in the analysis and it is impossible to have more than there are cases. The number should be kept far below this and certainly fewer than half. A successful analysis requires several hundred cases so the limitation is not serious. On the other hand, it is misleading to omit from the analysis a variable which, because of a correlation with one of the factors, affects the dependent variable through this missing proxy factor.

For example, suppose that the larger the floor area per storey the lower the cost per unit area and the greater the number of lifts required. If floor area per storey were missing from the factors in the analysis it could appear that cost per unit area fell as the number of lifts increased.

This effect of proxy factors is one of the reasons why relationships detected by multiple regression analysis cannot be assumed to be causal relationships. Association between two variables may or may not indicate that change in one causes change in the other.

8.6.4 Residuals

After a computer program has calculated a regression model, it offers the user the facility of examining the residuals. A *residual* is the difference between the value of the dependent variable for an item of data and the corresponding prediction of it made by using the regression formula. The residuals give a direct way of viewing the accuracy of the fit of the model to the data which have been used to calculate it. The sum of their squares is output by the computer program and is used to judge the significance of the fit.

A good way of studying the residuals is to plot them against the dependent variable and then against each of the independent variables in turn. A great deal can be learned from where the fit is poorest: it may suggest a transformation or another variable worth trying. At least it warns about the need to apply the model with extra caution in ranges of values of the variables where the relationships are far from linear.

Examination of the residuals plotted against each of the variables in turn may reveal one for which the assumption of linearity is clearly wrong but which cannot be dealt with by a simple transformation. This could easily be so in the formula quoted at the beginning of this section (8.6).

Cost per unit area = 12.95 − 0.002 × floor area + 0.38 number of storeys + 2.32 if in London

When the residuals are plotted against number of storeys it may be found that they are positive for 1 and 2 storeys, negative for 3 and 4 storeys and positive for 5 and more storeys, as in Fig. 8.3 in which a median fit has been drawn.

There are several ways of dealing with this. A curve fitting method would give a precise result for the particular sample, but the quality of the data would not justify an elaborate formula. However, it could form a useful starting point for the development of a simpler formula.

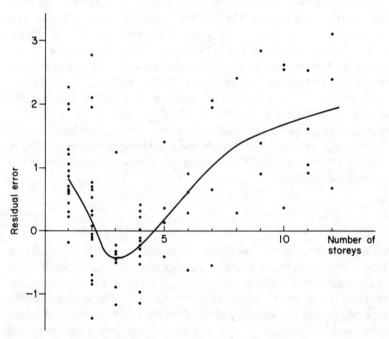

Fig. 8.3 The residual error in the estimate of the dependent variable cost per unit area.

Another way would be to devise two formulae: one for low rise and another for high rise buildings. The formulae would have different relationships for low and high rise which could be separately developed by trial and error.

For 1, 2 or 3 storeys the term for number of storeys could be

$$0.6 \times (2.3 - \text{number of storeys})$$

For 4 or more storeys it might be

$$2 \times \log_{10}(\text{number of storeys} - 3) - 0.2$$

The required coefficients could be finely tuned by writing a program to calculate residuals for the new formula and trying various coefficients until the plot in Fig. 8.3 became reasonably horizontal. The same program would also show whether any adjustment was needed to the constant in the main formula, at present 12.95, to make the average estimating error zero. The residual sum of the squares should be slightly less with the revised formula.

The same principle can be used to modify the regression model subjectively in any respect. There is no need for reluctance to do this. The assumptions underlying the model are so unlikely to be true and the quality of data so poor that modification by the application of judgement is more likely to do good than harm. In fact, although multiple regression analysis provides a good starting point it is possible to do without it and begin with a subjective formula. This can be repeatedly modified by trial and error using a specially written program and a micro-computer to calculate and examine residuals. The aim is to reduce residuals by successive modifications. It is important not to be carried away by enthusiasm for small improvements to the fit of the model to the sample, which may not be sufficiently representative to warrant such precision.

After an adjustment the residuals may acquire a new relationship to one of the variables or to a characteristic not previously in the analysis. The introduction of a new variable may be called for or the adjustment may be shown to be unwise.

8.6.5 Significance of the coefficients

An example of the procedure was included in a paper read by the author to the AGM of the Building Cost Information Service (see reference 2). If the standard error of a coefficient is less than half the size of the coefficient, it indicates that the lower 95% confidence limit is greater than zero. This shows that the factor to which the coefficient relates has some effect on the dependent variable. This is statistical significance. It says no more than that the effect in the sample is very unlikely to have been due to chance. The question remains, however, of whether such a coefficient is of any practical significance.

To decide whether the coefficients which have been output by a multiple regression analysis have any practical significance, it is helpful to insert in the regression formula produced by the analysis the mean values of the variables. These are usually part of the output. Multiplying by the coefficients gives the average effect of each factor on the dependent variable.

8.6.6 Dummy variables

It has been assumed so far that all the factors are numerical variables. In practice some will be descriptive, such as type of construction. Descriptive factors have to be dealt with by the device of turning them into a series of dummy variables, each of which can have only the values 0 or 1.

For example, instead of 'type of construction' being a variable, each type is treated as a separate variable. If the first type is load-bearing brick, every case in the analysis is coded 1 if it is of load-bearing brick construction and 0 if it is not. This produces a variable called 'load-bearing brick' with only two possible values. If the next type is pre-cast concrete every case is coded 1 if it is of pre-cast concrete and 0 if it is not. This produces another variable called 'pre-cast concrete', and so on. The coefficient which multiple regression analysis produces for a dummy variable can be regarded as the amount which must be added to the dependent variable if the descriptive factor applies. 'If in London' in the formula in Section 8.6 is a dummy variable. It is the sole survivor of a series of dummy variables for all the UK regions.

8.6.7 Stepwise regression

If all variables are included in the analysis, those whose coefficients are not statistically significant should be dropped and the analysis repeated. There is a version of multiple regression analysis which does this automatically. This is one type of stepwise regression, but other types exist. If such an automatic method of factor selection is used, its criteria must be studied. Its use can save trouble but control of factor selection must be maintained. Also, practical significance must be borne in mind as well as statistical significance, and no stepwise method can do this.

8.6.8 Population drift

The final warning for users of multiple regression analysis of building price data concerns population drift. To explain this it is necessary to mention the residual standard error which is printed near the end of a computer analysis. This should be considerably less than the starting standard deviation of the dependent variable. Like any standard error, it indicates the amount by which the estimate of the population parameter, in this case the arithmetic mean, may be 'wrong' by chance. About twice the residual standard error is usually taken to be the most that an estimate made using the regression formula could by chance be different from the 'best' estimate or population mean.

Thought must be given to the population from which the sample should be considered to have been drawn. When this is decided the stability of the population must be assessed. Changes in the population are continually occurring. Some changes affect the relationships which the analysis measures and some affect unconsidered variables not present in the regression formula.

As a very rough rule of thumb, the author's experience is that when any estimating formula is applied the errors experienced are at least 50% more than were forecast by the residual standard error even when no reason is known for population drift.

8.7 Design assumptions

Some of the data for any cost model, especially realistic simulations, are not available until late in the design stage. This raises for the client's cost adviser the most interesting and important feature of every proposed method of representing cost – the way in which assumptions about the design are to be made.

Assumptions should be as obvious as possible and easily changed by the user of a method. Some methods seem to make fewer assumptions about unknown data than others and are often favoured for this reason. The characteristic for which they are chosen is really their concealment of assumptions, and this is not in itself a virtue. This is not to say that methods in which assumptions are deeply embedded cannot be good for other reasons. It does, however, mean that, in all working methods, the user should discover what assumptions they make before taking them into use. He should try to keep the assumptions in mind when interpreting results.

Some methods provide such ease of use that there is reluctance to examine their assumptions. The decision to use them is unlikely to be affected and there is an unconscious desire not to learn anything which reduces confidence in the results of using them. For many users this is true of methods of costing which assume fixed relationships between quantities of materials and the resources required to put them in place.

Assumptions about an embryonic design should not be baulked or concealed. The designer may not like to see such assumptions being made but must understand that, whatever he intends, assumptions have to be made if he is to be given an estimate of cost. The choice is only between seeing them and hiding them. Even when an estimate is not being provided, advice on the effects on cost of alternative design decisions requires assumptions about the design. It is better to use a method which reveals them, at least to the cost adviser, than one which conceals them. In the long run, exposure of design assumptions should improve the dialogue and mutual understanding between the designer and the cost adviser.

Although there is much resistance to the principle of keeping a data bank of costed standard designs, it has great attraction from the aspect of assumptions. The assumptions implied are clear and the task of producing a construction plan or cost plan is reduced to consideration of the effects of changes from each of the nearest few standard designs. It is easier to keep track of the assumptions being made and it may make clearer the directions in which economies may be sought. It is better to work from two, three or more standard designs and compare results. If the reason for differences between them cannot be reliably determined, cost advice could be confined to respects in which two approximately agree. Estimates could be separately calculated and the median used.

The method of adapting costs for standard designs to produce estimates for another design depends on the way the costings have been made. Any sort of cost modelling technique can be used. The formulae produced by regression analysis can be used, unit prices for quantities or areas of elements can provide the link, or construction plans can be stored and adapted.

Reference

1. Morrison, N., and Stevens, S. (June 1981) *Cost Planning and Computers*, Department of the Environment (Property Services Agency).
2. Cost Study F29 for subscribers to the Building Cost Information Service of the Royal Institute of Chartered Surveyors (November 1982).

9 The accuracy of estimating

9.1 Measuring performance

Few practitioners objectively measure their estimating accuracy. When they do they are usually unpleasantly surprised. Because of the optimistic idea that most have of their own performance they too easily reject a new method for which the accuracy has been quantified. They find it hard to believe that they cannot do better than the figures quoted for the new method.

For example, a computer method which, wrongly as it happened, claimed to provide 60% of estimates within 10% of the lowest tender was treated with scorn by some practitioners as being uselessly inaccurate. In fact few could do better than the quoted performance. Unfortunately the computer method could not, in practice, do nearly as well as was claimed because the claim was based on its performance when applied to the few data which had been used to derive the method. It would not do nearly so well when applied to new data.

To measure estimating errors a definition of error must be decided upon. For the client's estimator the obvious one is the difference between the estimate and the lowest tender. Monitoring this has already been described in Chapter 7 where the difference was called D, and the D curve formed the basis of a way of improving bidding and estimating. Other targets may be aimed at but, whatever the target, a convenient way of expressing the error is the ratio of estimate to target.

The contractor's estimator may have more difficulty in defining his target and measuring his errors. He may aim to be in a particular place, such as second or third in the ranking of bids for a contract. In that case he would be allowing his estimating errors and those of his competitors to produce the required proportion of lowest tenders. Where bids are published he could compare his bid with that of the one in the target position, sometimes himself, and call the difference his error. He would tune his estimating method until his average error was zero. The method is a little clumsy because, if he is not satisfied with the proportion of successes even though his average is right, he will have to shift his aim to a different target.

The methods of Chapter 7 provide a more refined version of this strategy. If he is using that method he will be estimating his costs on a consistent basis before using the strategy to arrive at a bid. The accuracy of this cost estimate should be monitored by comparison with actual costs when they are known.

As the actual costs can only be known for contracts which are won, there is a danger of tending to judge the estimating method when it has performed successfully. On the other hand, it can be argued that the true measure of effectiveness of an estimating method is the profitability of contracts won, so only these are relevant. This is difficult to rationalize because if a different method had been used, different contracts would have been won, so comparison between methods could be biased.

One way of improving the comparison would be to calculate several bids based on different estimating methods and assess whether some would have avoided winning the less profitable contracts while retaining the more profitable. Even this can be misleading because some of the lost contracts which would have been won by the unsubmitted bids might also have had low profitability.

As well as measuring the error by comparison with the actual target, it may be interesting to compare the estimate with a more stable hypothetical target. Reducing the contribution of 'target movement' to the variability of the error brings us closer to measuring the inherent error of the estimate.

For example, the average of the bids for a contract is more stable than the lowest, so it is quite reasonable to aim at this and then make an adjustment for the difference between the average and the lowest, taking into account the number of tenders, as described in Chapter 7. An estimator who can show that the correlation between his estimates and average bids is good (i.e. the difference does not vary much) is entitled, in one sense, to claim good performance; even though it is little consolation to a client whose lowest tender has come in far above the estimate.

Because of occasional wildly erratic bids, the median is a preferable average to the arithmetic mean. Also, it is easily calculated because bids are likely to be arranged in order of size for presentation to the client.

The idea of measuring error in relation to various targets brings out the importance of distinguishing between average error, however measured, and dispersion of errors. The quality of a rifle is measured by clamping it rigidly and recording how closely together its shots fall; the size of the group. Quality is not judged by how much it needs zeroing to bring the centre of the group on to the centre of the target. In the same way, the quality of an estimator can best be judged by the dispersion of errors and not by the long-run average error, which can easily be corrected and may even be deliberate. Many estimators aim high because clients are happier to accept errors in that direction than the other.

Another reason for using the dispersion of errors about their average separately from the average error is that the effects on dispersion of choosing various defined points in the tender list as hypothetical, or initial, targets can be compared.

The dispersion of errors can be thought of as either the standard deviation of percentage errors or the standard deviation of the percentage ratio of estimate to target. They are the same. Of course, other measures of dispersion could be used.

9.2 Present achievement

Although slightly skewed to the right, the distribution of estimating errors can be regarded as conforming to the normal distribution. A typical standard deviation of percentage errors made by clients' estimators was found by Morrison and Stevens (see reference at end of chapter) to be 12%. Specializing in a particular type of work can reduce it to 10%, and for high-value buildings it is also about 10%. However, the author has seen no documented evidence of lower standard deviations than these figures measured over enough estimates to be convincing. Also, if sums for which estimates are not being made are excluded, the performance is much worse.

In this connection the standard error of the standard deviation must be remembered. It is $\sigma/\sqrt{(2n)}$, where σ is the population standard deviation and n the number of estimates. Therefore, if the population standard deviation is taken to be 12%, the number of estimates needs to be at least 50, giving a standard error of 1.2, before a sample standard deviation of 10% can be regarded as significantly better.

Contractors' estimators seem to do better but their task is different. It is not easy to interpret the meaning of their ability to provide a basis for bids whose dispersion has a coefficient of variation (c.v.) of 6%, but it is certain that their methods produce results which agree with each other more closely than with those of the client's estimator. This is not surprising because even if the client's estimator used better methods which were in some sense better than the contractors' he would still agree with them more closely if he changed to using their methods as far as possible. It does not logically follow that contractors' estimators' methods would provide a better basis for advice to clients and guidance to designers, but it does seem likely. One factor which could contribute to a better performance by contractors' estimators is that they take into account construction methods more realistically than clients' estimators. If so, this would certainly provide a means of improving advice to designers.

When contractors estimate their costs, their variability is presumably less than that exhibited when their mark-up variability is included. Therefore the variability of contractors' estimates of their costs is probably a little less than the variability between bids for the same contract. Allowing a coefficient of variation of 3% for variability of mark-up, and using the additiveness of squares for independent causes of variation, this leaves about 5% for the coefficient of variation of estimates of costs:

	c.v.	(c.v.)²
Total c.v. of bids	6%	36
− c.v. of mark-ups (as percentages of bids)	3%	9
c.v. of cost estimates	5%	27

Clients' estimators could reasonably hope to reduce the variability of their estimates to close to this level if they used the same methods as contractors when estimating their costs. To calculate the resulting estimating errors requires the addition of the variability of the winning contractor's mark-up, which is presumably very small. Winning bids are likely to have a less variable mark-up than other bids. If it has a c.v. of less than 2% and the c.v. of the errors introduced by lack of certainty about construction methods and other causes is less than 4%, the c.v. of the total client's estimating error becomes at most 7%:

	c.v.	(c.v.)²
c.v. of cost estimates	5%	27
c.v. of mark-ups of winning bids	2%	4
c.v. of other causes	4%	16
Total estimating error (client's)	7%	47

This would be a much better performance than present methods produce.

Clients should be made aware of the dispersion of estimating errors. The proportion likely to be within, say, 10% would be a suitable way of doing so. Also, the limits of estimating accuracy should be borne in mind when deciding whether to proceed with a design. It would be a pity to reject a good design which seems too expensive when it would be acceptable if the estimate were too high by about one standard deviation of the distribution of errors. Equally, the client should be prepared for the lowest tender to be the same amount or more above the estimate.

Being aware of the distribution of estimating errors also helps when deciding whether to accept a tender as the best obtainable or to call for more.

9.3 Improving estimating performance

Ways of improving estimating performance have been mentioned in previous chapters but they are brought together here and reviewed.

An improvement requiring no new techniques but more effort would be obtained by use of more than one project as a price analogue. Each estimate can be based on a single previous project but the estimating process should be repeated for as many other past projects as are available or there is time for. Even if each estimate has to be made less thoroughly, the median result is likely to be better than just the first one. It is more productive of accuracy to spend time on making repeated estimates than on refining any one of them.

It may be necessary to coarsen the description which defines an analogous project in order to obtain more than one. If so, this should be done even if adjustments to allow for differences have to be large. Of course, if they are wildly unreliable such a project cannot be used. A minimum of four projects should be the aim, but even two or three are much better than one.

Some time can be saved if an estimate being calculated is clearly going to be very high or very low compared with the others. It need not be accurately completed because its exact value will not affect the median.

Making several estimates and averaging them is a powerful way of improving long-term performance, because the standard error of the average is proportional to $1/\sqrt{n}$ provided that the estimates are independent. They will only be independent if each is based on a separate price analogue. If the same price analogue is used but different estimating methods are employed, the improvement will be very much less. However, it is better to use two good methods than one, perhaps weighting them before averaging as described in Chapter 3.

Another simple procedure to improve estimating performance is to use several methods for each estimate and to keep records of errors so as to select the best method or combination of methods. The methods will not all differ by much so the extra work entailed in using several is not necessarily great. Again, the standard deviation of percentage errors should be the basis of comparison and enough projects must be estimated for the comparison to be valid. As already shown, it will probably require over 100 to make fine comparisons but the worst can be eliminated after about 30. It is important not to draw hasty conclusions based on fewer comparisons. The t-test provides a formal statistical basis for conclusions.

Dispersions should be compared thoughtfully. A method which has the smallest dispersion of errors but on rare occasions can produce a very large error will not necessarily be preferred to one without this characteristic even if it has a rather greater dispersion. Occasional alarmingly large errors characterize oversimplified methods which, while working well for the great majority, fall down completely on one that is out of the ordinary in one or two crucial respects. Experience gained by experimenting with the method is required to detect the cases where it is inadequate.

Moving towards contractors' methods of estimating was suggested earlier as a way of improving clients' estimators' performance. This would include not only learning from them how to take account of construction methods (see Chapter 8) but also emulating their way of arriving at a mark-up. This would necessitate judging the market in the same way and following some aspects of a bidding strategy. A way of doing this was described in Chapter 7.

The amount of improvement in estimating accuracy which can be expected by adopting better methods is impossible to calculate with any certainty without experimentation. Such experiments would be valuable, but an idea of the improvement which it would be possible to make to clients' estimators' performance can be gained from studying the accuracy of contractors' estimating. This reasoning, earlier in the chapter, led to a hoped for coefficient of variation of clients' estimators' errors of 7%.

Whatever method of estimating is used, it will be improved by better data. With more estimators having access to computers, large data banks should become easier to handle. So far this is not yet accomplished but programs are being written and before long, data retrieval systems which are economical and suitable for building price data will be available. Some of the characteristics required of a data retrieval system and ways of handling the data in the bank were given in Chapter 3.

9.4 Helping the contractor

When the client's estimator has done all that he can to improve his own estimating methods, including using contractors' methods where possible, he has provided himself with the best possible 'rifle' for hitting the target. The target is moving partly because of the variability of the contractors' estimates, so he will reduce the movement if he can improve the contractors' estimates. The only way he can do this is to improve the information upon which the contractors base their tenders.

Research is required to identify which aspects of information need improving. For example, the coefficients of variation of tenders could be compared for designs in various stages of detail. It may be that incomplete working drawings increase the dispersion because the contractors make different assumptions about the missing detail.

Another cause of target movement is, of course, differences in keenness to win the contract. The best way to reduce these is to increase the attractiveness of projects so as to make the less keen tenderers keener. This will not only reduce variability but, more importantly, tend to reduce the lowest bid. This is because among the bottom few tenderers estimating errors are almost certainly greater than differences in mark-up, so the lowest tenderer may not be the keenest.

Some clients and designers may be unable to see that improving the ability of contractors to estimate accurately will, in the long run, be advantageous to the client. It will reduce the number of low bids resulting from large estimating errors but the general level will be lower, and less variable, because contractors will not need to include in their mark-up such a large safety factor to mitigate the worst effect of their estimating errors. Neither will there be so many loss-making contracts which inevitably give trouble and tend to lead to difficult claims.

9.5 Computer methods

The use of computers can improve the accuracy of estimating in two important respects. First, they can make available a large bank of data so estimates can be based on more projects. Second, they allow the consistent application of quite involved methods.

The advantage of the former is unequivocal. The latter is only as useful as the chosen methods. Consistency of method is helpful because it facilitates objective monitoring of performance, improves the development of the user's judgement and provides a basis for a bidding strategy. Elaborate methods can be good if the user thoroughly understands what is going on and can intervene in a way which allows his own judgement to be powerfully applied in the areas where it is required. The computer should do the hard work but leave the estimator in control.

Reference

Morrison, N., and Stevens, S. (June 1981) *Cost Planning and Computers*, Department of the Environment (Property Services Agency).

10 Forecasting

10.1 The need for forecasting

There are two main needs for forecasting in building. First, the future pattern of expenditure in a project or programme of projects has to be forecast so that funds may be efficiently managed. Programme expenditure forecasts can best be built up from forecasts for individual projects. Second, predictions of the inflation or deflation of building prices affect the management of funds for a building programme, and also for a single contract if payment is made for price fluctuations during its course.

10.2 Principles of forecasting

Good forecasting, like so much in the analysis of building price data, depends on disciplined blending of objective and subjective methods. They can be founded on analyses of past data or they can be assessed or modified from experience. It is important to record assessments and the reasoning behind them just as carefully as the results of analyses. In this way consistency can be improved, lessons can be learned and the forecasts refined as rapidly as possible in the light of experience. Recording assumptions underlying subjective methods also provides a vehicle for reasoned discussion and allows the experience of the old hand to be passed on in the best way.

10.3 Project expenditure forecasting

10.3.1 Construction programmes

If a construction programme is agreed before work begins, this provides a basis for forecasting project expenditure. First, the programme should be evaluated in terms of time and expenditure, then this can then be turned into a forecast. Construction programmes are generally optimistic, so records of past programmes should be analysed to provide correction factors. As well as the construction programme, another basis for forecasting is required for use before the programme is available or for projects for which a suitable one is not provided.

10.3.2 S-curves

A basis for forecasting the expenditure pattern in an individual project is provided by the familiar S-shaped curve derived from plotting the cumulative percentage of total expenditure against percentage of total time for as many projects as possible. Separate curves may be required for some types of building. If the only use for the curve were forecasting the expenditure pattern, 'total time' would be the duration the contracts had been planned to run. This is because planned time corresponds to the time which will be used when the expenditure forecast comes to be made. Actual duration will, of course, be unknown when forecasting. However, to use planned duration alone would limit the usefulness of the curve when monitoring progress so it is better to use actual duration when setting up the S-curves. Payments to date, as a percentage of total payments, are plotted against time elapsed to date as a percentage of total duration of construction. The relationship between actual duration and expected or planned duration can be the subject of a separate graph or formula which can be used to make a forecast of actual duration before using the expenditure curve to forecast the expenditure pattern.

Although the cumulative expenditure curve is usually described as S-shaped, the 'S' is such a long one that it is reasonable to approximate it by a long, steeply sloped straight line with two shallower slopes at the beginning and end. Figure 10.1 shows how good the approximation is. The advantage of doing this is that it is easier to represent the curve in a computer program. Three simple formulae are required instead of one elaborate, and not always satisfactorily representative, formula produced by a curve-fitting program.

To provide the S-curve it is necessary to analyse past data. First, the expenditure pattern for each project is adjusted according to recorded rules. The adjustments must be made without knowledge of whether the expenditure was fast or slow. The rules could be built up by examination of project records. Where a foreseeable sudden change in expenditure rate or a foreseeable delay or acceleration is recorded, the cause should be noted and an adjustment for that cause decided upon. As the adjustment rules are codified they can be refined by analysis of past records and applied to future projects for which forecasts are required. This is why the accelerations or delays must be foreseeable. Examples are the effect on expenditure of the installation of mechanical plant and the slowing-down effect of winter working at certain stages. As far as possible these effects should be quantified and used to adjust past records before plotting their S-curves. Later, the same methods of adjustment should be used, in reverse, for adjusting a forecast to allow for the special features of the project in hand.

Next, the cumulative expenditure curves are plotted for all the projects on the same graph or on transparent paper so that they can be overlaid. One set of curves relates to the adjusted data but a separate set is for the unadjusted data which will be required at the end of the forecasting process.

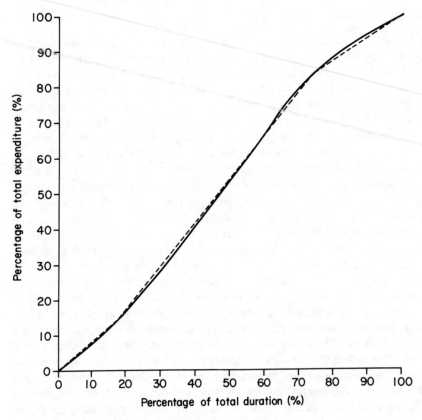

Fig. 10.1 A typical project cumulative expenditure curve.

It is interesting to compare the two sets of curves to see the overall effect of adjustments. They will probably have had little effect on the overall variability between the curves except that they may have moderated some outlying curves. It may be decided that some of the adjustments are not worth the trouble of making because the effect they represent will never be large. This requires careful thought because once adjustment procedures are decided upon it is important to keep to a consistent policy. A change would require revision of all the data upon which the curves are based.

The next step is to plot the median curve for the adjusted data, as in Fig. 10.2. This is done by marking, at each interval of the time axis, the point which equal numbers of curves are above and below. This median curve can be used for forecasting expenditure patterns.

Before basing a forecast on the median curve, it has to be adjusted by using, in reverse, the adjustment factors which had been applied to the data when

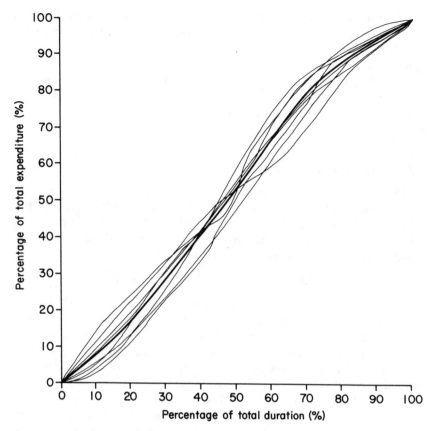

Fig. 10.2 Eight cumulative expenditure curves and their median.

producing the curves. This time, of course, they are related to the project in hand.

Finally, when planned construction duration has been converted to a forecast actual duration using the graph or formula mentioned earlier in this section and to be described shortly, the adjusted curve can be used to forecast the expenditure pattern. Early in the construction phase or before it starts, the adjusted curve can only be used as it stands. Once construction is well under way it can be used to forecast the completion date if the pattern of expenditure so far is continued; or the change in expenditure rate which would be required to complete construction in the planned time.

It is a matter for judgement based on the observed differences in past expenditure patterns, such as those in Fig. 10.2, how soon it is reasonable to start making forecasts of overrun or underrun of contract duration based on expenditure to date. For the first, say, 20% of expenditure it would be wise only to remark that initial expenditure is unusually low or high. Reasons should be sought but forecasts should wait until at least 20% of expenditure has been incurred. The family of past expenditure patterns should be examined to give an idea of how great the forecasting error could be. This can be done by calculating the forecast which would result from expenditure following the upper or lower quartile curve instead of the median curve. If there are more than about 20 curves the upper and lower deciles can be used as well.

10.3.3 Planned and actual duration

Analysis of past records produces the relationship between planned and actual duration. The latter is likely to average roughly 30% more than the former. Plotting one against the other provides points which can be represented by a smoothed median line, by a fitted curve or by a simple formula. A formula worth trying would be of the following type:

$$\log(\text{actual duration}) = D + C \log(\text{planned duration})$$

The simplest way of deriving it would be to plot actual against planned durations on log × log graph paper (that is, graph paper on which both scales are logarithmic) and drawing a median straight line through them. Then D would be the value at which the line cuts the log (actual duration) axis and C would be the slope of the line measured by the amount by which log (actual duration) increases for each unit increase in log (planned duration).

It is likely that the relationship will be affected by size of building so, if multiple regression analysis can be used to evaluate the coefficients, another term such as B log (contract value) or B log (floor area) should be added. If the facilities for multiple regression analysis are not available, the relationship between actual and planned duration should be evaluated for two or three size bands to provide three formulae to be used as appropriate. For projects whose size is close to the next size band, the formulae for both bands should be applied and the average answer used.

It would be wise to check whether the relationship holds for all types of buildings. It is possible that a complex technical building is more prone to delays than a simple one.

10.4 Forecasting price movements

The prices for which movements are to be forecast may be actual prices of key materials, wage rates or indices of prices or wages. Indices may be complex amalgams of prices intended to reflect movements of whole building prices. Whatever they are, the principles of forecasting them are the same.

10.4.1 Forecasting by analogy

An ideal method would link the price to be forecast to indices or other figures which its movements tend to follow. For example, movements in a tender price index ought to follow changes in costs of materials, plant and labour and the amount of work on the market. If this could be established and quantified, use could be made of the relationships. Forecasting would be best if movement of the tender price index lagged well behind those of costs and workload. If there were a correlation but no lag, forecasting would only be facilitated if changes in costs could be predicted from, perhaps, wholesale price indices and wage settlements and if the amount of work on the market could be predicted from the amount of work in architects' offices. Forecasting by such methods requires a short chain of correlations which includes a lag.

Correlations can be discovered by assembling monthly or quarterly data for at least 5 years and preferably much longer. The correlation coefficient between the price or index to be forecast and each of the possible indicators should be calculated with various lags and the correlation coefficients compared to select the best lag to use. The lag should not violate common sense.

The probability that a calculated correlation coefficient or a greater one could have arisen from chance fluctuations can be found from the following formula. It assumes normally distributed variables so any transformation of a variable which would reduce marked skewness should be made first.

$$t = \frac{r\sqrt{(n-2)}}{1-r^2}$$

where r is the calculated correlation coefficient, n is the number of pairs of numbers being correlated, and t is then looked up in Table C with $n-2$ degrees of freedom.

Correlations for which the probability that the effect could have arisen by chance is greater than about 0.2 should be rejected, but below that they should be given further consideration provided that the correlation coefficient itself is more than about 0.3. The effect should also be clear when the indicator and the price being forecast are plotted on transparent graph paper and slid along by the appropriate lag.

When correlations are being sought, many will be found which are strong but of no use for forecasting because both variables are moving under the influence of another variable. It is only the correlation of price with this other variable which is of interest, because it is a causal relationship. For example, prices and many indices tend to follow the general inflationary trend in the economy. Relationships between prices and these indices have a predictive value no better than an assumption that prices will follow the general inflationary trend. Predictions of this trend are widely available. Although they are unreliable they may provide the best available foundation for forecasting price movements. Attention can then be focused on reasons why building price movements are likely to differ from the general trend.

10.4.2 Detecting a seasonal pattern

If there are cyclic movements present in any of the variables, including the one to be forecast, correlations will be affected. It may be that the correlation is entirely due to both variables moving in the same seasonal pattern. Such a correlation may be interesting but is unhelpful, and before further tests this cyclic movement should be eliminated. Even if the seasonal effects are not entirely responsible for a correlation, a variable becomes a better predictor if seasonal movements are treated separately. The principle of removal of cyclic variation is to evaluate the average cyclic effect and then to adjust all data for this average effect.

Most cyclic variations repeat each year, reflecting a seasonal pattern which is intuitively reasonable. If shorter or longer cycles are detected they should be regarded with great doubt unless a reasonable mechanism for them is clear. It is a weakness of the methods of time series analysis that they often throw up patterns in the data which, though not due to chance, have no predictive value because they are no more likely to be repeated in the future than any other pattern. Time series analysis has become overdeveloped because of the intense interest in forecasting economic and financial indicators even if the improvement on simple methods is slight. Although most of the elaborate techniques of time series analysis and forecasting are inappropriate to building price data, evaluation of seasonal effects is simple and is unarguably justified for many time series. Regression methods are sometimes used to establish a trend line from which seasonal departures can be measured, but the assumption of an underlying straight line is seldom justified. A more satisfactory method is to study departures from the annual moving average. The first problem is to ensure that the moving average is correctly 'centred'. The example in Table 10.1 makes the method clear.

Quarterly data will be assumed, so a four-point moving average will be appropriate. If data are monthly, a 12-point moving average will be required.

Recentring by averaging adjacent values is to bring the effective dates of the moving averages into coincidence with the dates of the original values. It is only required where the number of data in the moving average is even; in this case 4.

Departures of the indices from the moving averages are calculated as percentages, with the sign showing whether the departure was upward or downward. The average of these departures for the first quarter of every year is the seasonal effect of the first quarter. Similarly, the effects of each of the other three quarters are calculated and the figures for each quarter are brought together as in Table 10.2.

Building tender prices are subject to sudden changes. In a short series, the

Table 10.1 Public sector building tender price index

Year	Quarter	Index*	Four-quarter moving average	Recentred moving average	Departure of index (%)
1976	1	102			
	2	104			
	3	110	106.25	107.50	+2.3
	4	109	108.75	110.62	−1.5
1977	1	112	112.50	114.12	−1.9
	2	119	115.75	118.12	+0.7
	3	123	120.50	122.75	+0.2
	4	128	125.00	127.38	+0.5
1978	1	130	129.75	132.88	−2.2
	2	138	136.00	138.25	−0.2
	3	148	140.50	144.12	+2.7
	4	146	147.75	151.38	−3.6
1979	1	159	155.00	159.25	−0.2
	2	167	163.50	169.25	−1.3
	3	182	175.00	180.12	+1.0
	4	192	185.25	191.00	+0.5
1980	1	200	196.75	199.50	+0.2
	2	213	202.25	203.50	+4.7
	3	204	204.75		
	4	202			

* Reproduced by kind permission of the Director of Quantity Surveying Services at the Property Services Agency of the Department of Environment.

Table 10.2 Percentage departures of indices from their moving average

	Quarter			
Year	1	2	3	4
1976			+2.3	−1.5
1977	−1.9	+0.7	+0.2	+0.5
1978	−2.2	−0.2	+2.7	−3.6
1979	−0.2	−1.3	+1.0	+0.5
1980	+0.2	+4.7		
Median	−1.0	+0.2	+1.6	−0.5
Range	2.4	6.0	2.5	4.1
Estimated population s.d.*	1.2	2.9	1.2	2.0
Standard error of median†	0.8	1.8	0.8	1.2

* Estimated from the range using Table B.
† Obtained from the formula $1.25\ \text{s.d.}/\sqrt{n}$, for a sample of size n.

quarter in which one of these happened would dominate the arithmetic mean for that quarter and affect the calculation of a mean seasonal effect. An example is the high figure for the second quarter of 1980 in Table 10.2. For this reason the median is a preferable measure for the average.

The dispersion of the departures is of interest because it gives a guide to how variable the data will be after adjustment for the average seasonal effect. It also measures the consistency of the seasonal effect and thus how valuable the adjustments are likely to be. The averages of the departures need not be much larger than their standard errors to provide evidence for the existence of a genuine seasonal effect, but even this requires several years' data. Also,

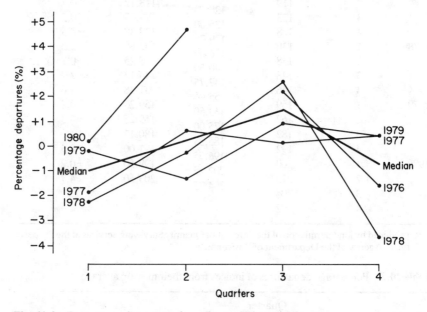

Fig. 10.3 Percentage departures from the annual moving average.

evidence from at least 5 years is needed to ensure that any effect detected has lasted long enough to indicate a persistent pattern.

It is instructive to plot a graph of the percentage departures in Table 10.2. Figure 10.3 does this, with the median marked with a thicker line.

There are good reasons for believing that in some parts of the public sector the uneven availability of funds during the financial year could affect tender prices. There is an apparent slight tendency for prices to be high in the third quarter and low in the first. In the third quarter the median is twice its standard error, so the value of t is 2.0 with 3 degrees of freedom. Table C shows the probability that such a difference from zero, or a greater one, could occur by chance. It is between 0.1 and 0.2. This is suggestive but not strong evidence. However, it is more likely to be a true effect than not, even though the evaluation of it is poor. If no better evidence is available it would be reasonable to assume a seasonal effect and to use this evaluation until more evidence accumulates.

10.4.3 Seasonal adjustment

If it is decided to use the existing data to make seasonal adjustments to the past and future indices, the average departures from the moving average will be used to calculate a seasonally adjusted index.

To convert the average (i.e. median) departures calculated in Table 10.2 to adjustments, their signs are changed before they are applied to the original data, as in Table 10.3. The adjustment for the first quarter is added to each of the indices for the first quarters, and so on for the other quarters. The resulting figures are seasonally adjusted.

Seasonal adjustment not only helps to prepare the data for forecasting but also enables underlying trends to be discerned more easily. This is generally

Table 10.3 Seasonally adjusted index
(From Table 10.2 the adjustments for the four quarters are +1.0%, −0.2%, −1.6%, +0.5%)

Year	Quarter	Unadjusted index	Adjusted index
1976	1	102	103
	2	104	104
	3	110	108
	4	109	110
1977	1	112	113
	2	119	119
	3	123	121
	4	128	129
1978	1	130	131
	2	138	138
	3	148	146
	4	146	147
1979	1	159	161
	2	167	167
	3	182	179
	4	192	193
1980	1	200	202
	2	213	213
	3	204	201
	4	202	203

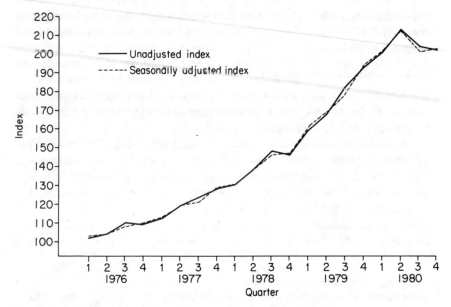

Fig. 10.4 The effect of seasonal adjustment.

understood to be necessary for unemployment figures but is less often regarded as important in other fields. It is possible for an index to fall but, when the seasonal effect is removed, the real tendency may be seen to be upward. Figure 10.4 shows examples of this between quarters 3 and 4 of both 1976 and 1978.

10.4.4 Smoothing

Some smoothing is achieved, incidentally, by seasonal adjustment. It is often recommended that the next step should be to smooth out minor fluctuations so that the general direction of movement can be discerned and, if desired, projected forward to provide the basis for a forecast. It is shown later that this is not always a good way of forecasting, but smoothing may help understanding of past movements and for this reason it is described here. The moving average is an unsuitable method of smoothing because it changes suddenly when an early high or low figure drops out. The most appropriate technique, exponential smoothing, has already been described in Section 3.4.3. It is a type of moving average but has the advantage that, instead of dropping out suddenly, early data have gradually less effect as time goes on, and the speed of response to change can be controlled by the choice of weights. The weights which achieve a suitable degree of smoothing can be found by trial and error. Only enough weight should be given to early data to provide the degree of smoothing necessary to expose trends clearly. Too much reduces responsiveness to movement; then changes in general direction are revealed too late.

As a general guide, weights of 0.5 on past data and 0.5 on the new item have been found to be about right for a price index, and they have the advantage of ease of calculation. They have been used in Table 10.4. For some indices other weights may be required. A more erratic index may, for instance, need weights of 0.7 and 0.3 to reveal trends. Weights of 0.5 and 0.5 should be looked upon only as a starting point in a series of tests of a forecasting method.

For smoothing the trend to enable an age correction to be made, more weight on past data may be required. Weights of 0.9 and 0.1 have been used in Table 10.4, but for comparison weights of 0.5 and 0.5 have also been used.

Table 10.4 Exponential smoothing of the seasonally adjusted index

Year	Quarter	Seasonally adjusted index	Exponentially smoothed seasonally adjusted index Weights 0.5,0.5	Trend	Exponentially smoothed trend Weights 0.9,0.1	0.5,0.5	Exponentially smoothed seasonally adjusted index with trend correction* Trend weights 0.9,0.1	0.5,0.5
1976	1	103						
	2	104	103.5					
	3	108	105.8	2.3				
	4	110	107.9	2.1	2.3	2.2	110	110
1977	1	113	110.4	2.5	2.3	2.4	113	113
	2	119	114.7	4.3	2.5	3.3	117	118
	3	121	117.9	3.2	2.6	3.3	120	121
	4	129	123.4	5.5	2.9	4.4	126	128
1978	1	131	127.2	3.8	3.0	4.1	130	131
	2	138	132.6	5.4	3.2	4.7	136	137
	3	146	139.3	6.7	3.6	5.7	143	145
	4	147	143.2	3.9	3.6	4.8	147	148
1979	1	161	152.1	8.9	4.1	6.9	156	159
	2	167	159.5	7.4	4.4	7.1	164	167
	3	179	169.3	9.8	5.0	8.5	174	178
	4	193	181.1	11.8	5.7	10.1	187	191
1980	1	202	191.6	10.5	6.1	10.3	198	202
	2	213	202.3	10.7	6.6	10.5	208	213
	3	201	201.6	−0.7	5.9	4.9	207	206
	4	203	202.3	0.7	5.4	2.8	208	205

* Trend correction is exponentially smoothed seasonally adjusted index +0.5/0.5 × (exponentially smoothed trend).

Figure 10.5 shows that exponential smoothing, using weights of 0.9,0.1 in the trend correction, had the desired effect except that, because of the accelerating trend, the attempt to compensate for the age of the smoothed indices by using age corrections based on a weight of 0.9 on past data, produced age corrections which were themselves out of date. They were too small and left the final smoothed index below its proper level. Smoothing the trend with weights of 0.5 and 0.5 improves the age correction but reduces the smoothing effect.

If logarithms of the seasonally adjusted indices had been used at the

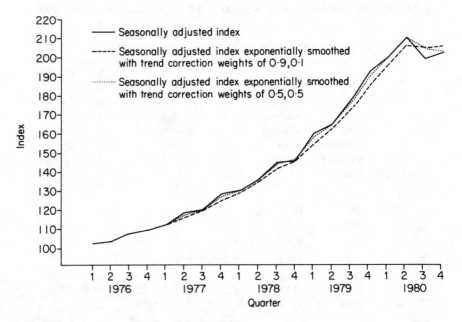

Fig. 10.5 The effect of exponential smoothing.

beginning, and all processes carried out in the same way as before, the trend would have been a little more steady and the age of the trend correction slightly less important. However, when the inflation rate of the index is increasing, as it is in this case, the 0.9,0.1 trend correction still produces an index which tends to be too low.

Although, if the trend is steady, exponential smoothing can provide a satisfactory basis for illustrating the trend, it does not always provide a good springboard for projecting it. The decision on the usefulness of the technique in forecasting is for the reader to make after studying Fig. 10.5. It is worth learning, in any case, because it is valuable for stabilizing the weights used in producing a composite index as described in Section 3.4.3.

10.4.5 Adjustment for inflation

The most obvious reason for a rise in a price index is general economic inflation. There is much to be said for dividing by an index of inflation, such as the retail price index, to adjust for this. If this is done, and average seasonal effects are taken out before forecasting is attempted, the search for reasons for movements becomes directed towards reasons why they should be different from the general inflation in the economy.

It is sometimes suggested that, instead of using an external index of inflation, the adjustment for inflation can be evaluated from the smoothed seasonally adjusted data themselves. This is useful for research into past data but, unlike seasonal variation, trends cannot be assumed to continue. Also, movements which have nothing to do with inflation are included in the evaluation. The method is unsuitable for forecasting.

10.4.6 Other adjustments

Provided that careful records are kept of the reasoning used, subjective adjustments can be made to the seasonally adjusted data to allow for any causes of price movement which the user believes he understands. As already described, a record should be built up of causes and the methods of quantifying them, however subjective they are, so that they can be applied to the data consistently. The same record will allow those which are appropriate to a forecast to be applied in reverse after the forecast has been made.

Making some of these adjustments is more difficult once exponential smoothing has been done, and this is another reason for not using the technique in this way.

10.4.7 Forecasting by projection and adjustment

Attention can now be turned to forecasting. A crude basis is that the future will be like the past; either that there will be a continuation of the existing trend or, if seasonal and inflationary effects have been temporarily removed from the trend, that there will be no change. This gives a basic forecast to which adjustments can be made corresponding, but in the opposite direction, to those made to the past data. Again use can be made of correlations with other data and of subjective assessments of likely influences. Again these should be quantitatively recorded and their net effect obtained. The adjustments made for general economic inflation and the seasonal effect would be among the adjustments made in reverse to give the final forecast.

When the true figure becomes known, it should be compared with the forecast. The record of assessed influences should be reviewed in the light of the forecasting error and lessons recorded for use in future adjustments and forecasts.

For thorough testing, the entire method of forecasting should be worked out and applied to the historic data to judge its performance. 'Forecasts' should be made at each point in the series and compared with the next item of data. Different methods and smoothing weights can be tried until the best combination of forecasting method and smoothing weights is found.

This procedure is a rationalization of what many people do intuitively. It is a far cry from the search for patterns and rhythms in past data and their forward projection, which is the obsession of some statisticians. The identification and evaluation of patterns is only useful if reasons can be reliably attached to them and the reasons reviewed for continued applicability.

Tables

Table A The proportions of area under the normal curve more than various numbers of standard deviations above the arithmetic mean. (i.e. lying to the right, as in Fig. 4.5)

Number of s.d.s above mean	Area to the right	Number of s.d.s above mean	Area to the right
0.0	0.500	1.5	0.067
0.1	0.460	1.6	0.055
0.2	0.421	1.7	0.045
0.3	0.382	1.8	0.036
0.4	0.345	1.9	0.029
0.5	0.309	2.0	0.023
0.6	0.274	2.1	0.018
0.7	0.242	2.2	0.014
0.8	0.212	2.3	0.011
0.9	0.184	2.4	0.008
1.0	0.159	2.5	0.006
1.1	0.136	2.6	0.005
1.2	0.115	2.7	0.004
1.3	0.097	2.8	0.003
1.4	0.081	2.9	0.002

Table B Range as a proportion of population standard deviation in samples from a normal distribution.

Number in sample	Average range ÷ standard deviation	Number in sample	Average range ÷ standard deviation
2	1.13	10	3.08
3	1.69	11	3.17
4	2.06	12	3.26
5	2.33	13	3.34
6	2.53	14	3.41
7	2.70	15	3.47
8	2.85	20	3.73
9	2.97	30	4.09

Table C Table of percentage points of Student's *t*-distribution

Double-tailed probabilities	0.50	0.40	0.30	0.20	0.10	0.05	0.02	0.01	0.005	0.001	
Single-tailed probabilities	0.25	0.20	0.15	0.10	0.05	0.025	0.010	0.005	0.025	0.0005	
Degrees of freedom											n
1	1.000	1.376	1.963	3.078	6.314	12.106	31.821	63.657	127.320	636.619	1
2	0.816	1.061	1.386	1.886	2.920	4.303	6.965	9.925	14.089	31.598	2
3	0.765	0.978	1.250	1.638	2.363	3.182	4.541	5.841	7.453	12.941	3
4	0.741	0.941	1.190	1.533	2.132	2.776	3.737	4.604	5.598	8.610	4
5	0.727	0.920	1.156	1.476	2.015	2.571	3.365	4.032	4.773	6.859	5
6	0.718	0.906	1.134	1.440	1.934	2.447	3.143	3.707	4.317	5.959	6
7	0.771	0.896	1.119	1.415	1.895	2.365	2.998	3.449	4.029	5.405	7
8	0.706	0.889	1.108	1.397	1.860	2.306	2.896	3.355	3.832	5.041	8
9	0.703	0.883	1.100	1.383	1.833	2.262	2.821	3.250	3.690	4.781	9
10	0.700	0.879	1.093	1.372	1.812	2.228	2.764	3.169	3.581	4.587	10
11	0.697	0.876	1.088	1.363	1.796	2.201	2.718	3.106	3.497	4.437	11
12	0.695	0.873	1.083	1.356	1.782	2.179	2.681	3.055	3.428	4.318	12
13	0.694	0.870	1.079	1.350	1.771	2.160	2.650	3.012	3.372	4.221	13
14	0.692	0.868	1.076	1.345	1.761	2.145	2.624	2.977	3.326	4.140	14
15	0.691	0.866	1.074	1.341	1.753	2.131	2.602	2.947	3.286	4.073	15
16	0.690	0.865	1.071	1.337	1.746	2.120	2.583	2.921	3.252	4.015	16
17	0.689	0.863	1.069	1.333	1.740	2.110	2.567	2.898	3.222	3.965	17
18	0.688	0.862	1.067	1.330	1.734	2.101	2.552	2.878	3.197	3.922	18
19	0.688	0.861	1.066	1.328	1.729	2.093	2.539	2.861	3.174	3.883	19
20	0.687	0.860	1.064	1.325	1.725	2.086	2.528	2.845	3.153	3.850	20
21	0.686	0.859	1.063	1.323	1.721	2.080	2.518	2.831	3.135	3.819	21
22	0.686	0.858	1.061	1.321	1.717	2.074	2.508	2.819	3.119	3.792	22
23	0.685	0.858	1.060	1.319	1.714	2.069	2.500	2.807	3.104	3.767	23
24	0.685	0.857	1.059	1.318	1.711	2.064	2.490	2.797	3.090	3.745	24
25	0.684	0.856	1.058	1.316	1.708	2.060	2.485	2.787	3.078	3.725	25
26	0.684	0.856	1.058	1.315	1.706	2.056	2.479	2.779	3.067	3.707	26
27	0.684	0.855	1.057	1.314	1.703	2.052	2.473	2.771	3.056	3.690	27
28	0.683	0.855	1.056	1.313	1.701	2.048	2.467	2.763	3.047	3.674	28
29	0.683	0.854	1.055	1.311	1.699	2.045	2.462	2.756	3.038	3.659	29
30	0.683	0.854	1.055	1.310	1.697	2.042	2.457	2.750	3.030	3.646	30
40	0.681	0.851	1.050	1.303	1.684	2.021	2.423	2.704	2.971	3.551	40
60	0.679	0.848	1.046	1.296	1.671	2.000	2.390	2.660	2.915	3.460	60
120	0.677	0.845	1.041	1.289	1.658	1.980	2.358	2.617	2.860	3.373	120
∞	0.674	0.842	1.036	1.282	1.645	1.960	2.326	2.576	2.807	3.291	∞

Table D The values of χ^2 with various probabilities of occurring by chance

Degrees of freedom	Probability (as a percentage)					
	50%	30%	20%	10%	5%	1%
1	0.5	1.1	1.6	2.7	3.8	6.6
2	1.4	2.4	3.2	4.6	6.0	9.2
3	2.4	3.7	4.6	6.3	7.8	11.3
4	3.4	4.9	6.0	7.8	9.5	13.3
5	4.4	6.1	7.3	9.2	11.1	15.1
6	5.3	7.2	8.6	10.6	12.6	16.8
7	6.3	8.4	9.8	12.0	14.1	18.5
8	7.3	9.5	11.0	13.4	15.5	20.1
9	8.3	10.7	12.2	14.7	16.9	21.7
10	9.3	11.8	13.4	16.0	18.3	23.2
11	10.3	12.9	14.6	17.3	19.7	24.7
12	11.3	14.0	15.8	18.5	21.0	26.2
13	12.3	15.1	17.0	19.8	22.4	27.7
14	13.3	16.2	18.2	21.1	23.7	29.1
15	14.3	17.3	19.3	22.3	25.0	30.6
16	15.3	18.4	20.5	23.5	26.3	32.0
17	16.3	19.5	21.6	24.8	27.6	33.4
18	17.3	20.6	22.8	26.0	28.9	34.8
19	18.3	21.7	23.9	27.2	30.1	36.2
20	19.3	22.8	25.0	28.4	31.4	37.6
21	20.3	23.9	26.2	29.6	32.7	38.9
22	21.3	24.9	27.3	30.8	33.9	40.3
23	22.3	26.0	28.4	32.0	35.2	41.6
24	23.3	27.1	29.6	33.2	36.4	43.0
25	24.3	28.2	30.7	34.4	37.7	44.3
26	25.3	29.2	31.8	35.6	38.9	45.6
27	26.3	30.3	32.9	36.7	40.1	47.0
28	27.3	31.4	34.0	37.9	41.3	48.3
29	28.3	32.5	35.1	39.1	42.6	49.6
30	29.3	33.5	36.2	40.3	43.8	50.9

Example The table shows that, with 2 degrees of freedom, there is a 10% probability that a value of χ^2 of 4.6 or more could occur by chance.

Further reading

This book does not cover more statistical techniques than are required by most practitioners in the field of building prices; neither does it deal with the underlying theory. If it has stimulated the reader to want to gain a deeper understanding and wider knowledge of statistical methods the books listed below are recommended.

Kalton, Bradford Hill and Moroney are easy to read. The first two are basic but Moroney goes as far as most will need, except for multiple regression analysis. Chatfield gives more mathematical theory and Wetherill more still. Chatfield touches on multiple regression analysis, but Wetherill covers it in more detail.

The book by Draper and Smith is comprehensive and will be sufficient for anyone carrying out multiple regression analysis.

Kalton, G. (1966) *Introduction to Statistical Ideas for Social Scientists*, Chapman and Hall, London.
Hill, A. Bradford, *Principles of Medical Statistics*, The Lancet Ltd, London.
Moroney, M. J. (1966) *Facts from Figures*, Penguin Books, Harmondsworth.
Chatfield, C. (1978) *Statistics for Technology*, 2nd edition, Chapman and Hall, London.
Wetherill, G. Barrie (1982) *Elementary Statistical Methods*, 3rd edition, Chapman and Hall, London.
Draper, N. and Smith, H. (1966) *Applied Regression Analysis*, Wiley-Interscience, New York.

For a thorough explanation of indices:
Tysoe, B. A. (1981) *Construction Cost and Price Indices: Description and Use*, E. & F. N. Spon, London.

Index

Accuracy of estimating, 147–53
Adjustment of data, 37
Adjustment, subjective, 38
Age correction, 33
Analogue, price, 6, 150
Analysis of variance, 20, 21
Appraisal, rapid, 97–105
Appraising data, 83
Area related methods, 134–8
Arithmetic mean, 89–92
Assumptions, design, 145
Average, 89

Banks
    data, 41, 48, 128–30, 132–3, 138
    network, 129–30
Bar charts, linked, 129
Bessel's correction, 15, 66–8
Bidding strategies, 114–26
Black box methods, 127
Blending, 39, 68

Calculators, 2
Cards
    optical coincidence, 44
    edge punched, 42–3
    captive punched, 43
    machine operated punched, 45
Cell width, 86, 104
Central limit theorem, 66
Charts
    quality control, 105–8
    cusum, 108–9
Chi-squared (χ^2), 58
Chi-squared test, 58–63
COCO (cost of contractors' operations), 128–9
Coding, 47
Coefficient of variation, 15
Comparison of two samples, 60–2

Computer bureaux, 4
Computers, 3, 4, 45, 152, 153
Confidence limits, 64, 71
    for the arithmetic mean, 67–8
    asymmetrical, 71, 72
    for the median, 69–70, 71
    single-sided, 72
Continuity correction, 63
Control charts, 105–8
Correction
    age, 33, 165
    Bessel's, 66–8
    finite population, 68
    Sheppard's, 89
Correlations
    in adjustment factors, 137–8
    in regression, 140
Cost control, 135
Cost limits, 51
Cost models, 127–46
Cost planning, 134–5
Cover prices, 112
Curve fitting, 137
Cusum charts, 108–9
Cyclic variation, 160

D curve, 115
Data appraisal, 83
Data banks, 41, 48, 128–30, 132–3, 138
Data structure, 46
Data, wild, 96
Deciles, 17
Degrees of freedom, 60, 62
Deviate, random normal, 130–1
Deviation
    mean, 12–13
    standard, 13–15
Distributions, 49
    normal, 54–8, 110
    standardized normal, 54–5

Dummy variables in regression, 144

Error, standard, 66
 of the mean, 66
 of the median, 69–70
 residual (in regression), 144
Estimating, accuracy of, 147
Expenditure curve, 155
Expert systems, 128
Exponential smoothing, 31–3, 118, 164–5

Factors, price, 30
Finite population correction, 68
Forecasting, 154
Forecasting price movements, 158–68
Forecasting project expenditure, 154–8
Freedom, degrees of, 60, 62
Frequency tables, 85, 88

Geometric mean, 92

Harmonic mean, 97
Histograms, 86–8
Hypothesis, null, 74, 78

Independence, 130, 131
Indices, 28
 costs, 35–7
 price, 29–35, 48
 unweighted, 30
 value-weighted, 30
In-place materials methods, 131–4
Interquartile range (IQR), 17

Judgement, 1, 3, 65, 68

Laspeyre type index, 30
Limits, confidence, *see* Confidence limits
Linearity in regression, 140
Location, measures of, 89–97
Logarithmic plotting, 34
Logarithmic transformation, 34, 52
Logarithms, 2, 95

Macro-climate, 117–18
Mann–Whitney test, 80–1
Mean
 arithmetic, 89–92
 geometric, 92
 harmonic, 97
 trimmed, 91
 weighted, 95

Median, 51, 92
 confidence limits for, 69–71
 standard error of, 69–70
 weighted, 96
Median fit, 136
Median regression, 139
Micro-climate, 117
Micro-computers, 4
Mode, 93
Model, cost, 127–46
Monte Carlo method, 130–1
Multiple regression analysis, 20, 38, 139

Nelson's Column, 58
Network bank, 129–30
Normal deviate, random, 130–1
Normal distribution, 54–8, 110
 standardized, 54–5
 tables of, 55
Non-parametric tests, 80–2
Null hypothesis, 74, 78

Paasche type index, 30
Parameters, 83
Percentiles, 16
Planning, cost, 134–5
Population drift, 144
Population, finite, correction for, 68
Populations, 8, 11
Price analogues, 6, 132, 150
Price books, 6
Prices, unit, 131–2
Pseudo-random sampling, 26–7

Quantiles, 16–18
Quartiles, 16–18

Random normal deviate, 130–1
Random number generator, 4, 27
Random sampling, 26–28
Random variability, 6
Range, 11
 interquartile, 17
Ranks, 80
Rapid appraisal, 97–105
Realistic methods, 127–31
Regression analysis, multiple, 4, 20, 38, 139
 corrections, 141
 descriptive factors, 144
 dummy variables, 144
 linearity, 140
 median, 139–40

residual standard error, 144
residuals, 141–3
stepwise, 144

Sample size, 72–4
Samples, 8
   stratified, 9
Sampling, pseudo-random, 26–7
Sampling, random, 26–8
S-curves, 155
Seasonal adjustment, 163
Seasonal effects, 160
Selection of data, 24–8
Sheppard's correction, 89
Sigma (σ), 15
Significance of regression coefficients, 143
Significance testing, 74–82
Simulation, computer, 128
Skewness, 49
Smoothing, 94, 164
   exponential, 31–3, 118, 164–5
Software, statistical, 3
Standard basket, 30, 31
Standard deviation, 13–15
Standard error, 66
   of the mean, 66
   of the median, 69–70
   of other parameters, 70–1
   residual (in regression), 144
Stepwise regression, 144
Stratification, 9
Student's t test, 75–80

Systems, expert, 128

Tally marks, 86
Tenders, number of, 118–20
Tender patterns, 110–14
Test
   Mann–Whitney, 80–1
   Student's t, 75–80
   non-parametric, 80–2
Testing significance, 74–82
Time-sharing computer bureaux, 4
Theorem, central limit, 66
Transformation, 52–4, 140–1
   logarithmic, 34, 52
   square root, 54
Trimmed mean, 91
t test, Student's, 75–80

Unit prices, 131–2

Variability, 5–23
   causes, 20–3
   random, 6–7
   unassigned, 7
Variance, analysis of, 20, 21
Variation, coefficient of, 15–16

Weighted mean, 95
Weighted median, 96
Weighting data, 40
Width, cell, 86, 104
Wild data, 11, 16, 96, 97